JOHN WILKINSON

KING OF THE IRONMASTERS

JOHN WILKINSON

KING OF THE IRONMASTERS

FRANK DAWSON
EDITED BY DAVID LAKE

The
History
Press

First published 2012

The History Press
The Mill, Brimscombe Port
Stroud, Gloucestershire, GL5 2QG
www.thehistorypress.co.uk

British Library Cataloguing in Publication Data.
A catalogue record for this book is available from the British Library.

ISBN 978 0 7524 6481 7
Typesetting and origination by The History Press
Printed in the EU for The History Press.

CONTENTS

FRANK DAWSON – A BRIEF CV

In 1978, with an arts degree from the Open University, a diploma in education from the University of Leeds and twenty years' teaching experience in the UK and Africa, Frank Dawson went to live and work at Castle Head, the eighteenth-century home of John Wilkinson, 'King of the Ironmasters'. He and a group of friends had acquired the property to establish there a private, short-stay residential field centre for studies by teenage and adult students. At the start Frank knew nothing of the Wilkinsons, but folk memories of their activities in the area led to documentary research into their lives and fortunes, and then to short study courses and field excursions, which he taught and directed. Annually, for twelve years, Frank gave a public lecture at Castle Head Field Centre on some aspect of the Wilkinsons' lives. He retired in 1997 when the field centre became part of the Field Studies Council. Since then, he has linked together a continuous story of the Wilkinsons, using further evidence gathered from private and public archives up and down the country.

FOREWORD

John Wilkinson was the important third man in the firm of Boulton and Watt, though he was never a properly constituted business partner. His acknowledged iron-making expertise and his engineering skills complemented James Watt's inventive genius and Matthew Boulton's entrepreneurial flair. He made the iron parts for the early Watt steam engines, suggested working modifications, promoted sales and organised transport. In the ten years from 1775, the three men were central figures in the dramatically developing industrial Britain.

Against this backdrop, documentary sources reveal a Wilkinson family drama on an epic scale: a father with a touch of genius, bitter quarrels between father and sons, the loss of beloved women in the uncertainties of childbirth, and a family in constant dealings with the personalities and events of the Industrial Revolution. Notably: the Darbys of Coalbrookdale; Richard and William Reynolds; Josiah Wedgwood; Joseph Priestley (who married a Wilkinson); and Samuel More of the Society for the Encouragement of Arts, Manufactures and Commerce, John Wilkinson's lifelong friend. These power relationships are closely examined in the building of the great Iron Bridge over the Severn; in the litigation involving Watt's patent; in some early industrial espionage involving the manufacture of cannon for the British Navy; and the Wilkinsons' contact with France when she was at war with England.

Everyone has heard of Boulton and Watt; few know of John Wilkinson's importance in their story. He created a vast industrial empire but had no son to inherit it, and his need of an heir led to a reputation in his old age as a womaniser and lecher. He quarrelled with some of his partners and with some of his family. Moreover he did not regard himself as bound by all the established conventions of the time: he consorted openly with a mistress, had three children by her in his seventies and left his vast empire to them, only for it to be consumed by litigation.

His contemporaries dubbed him 'Iron-mad Wilkinson'. It is a sobriquet at once patronising and dismissive. John Wilkinson rose from humble beginnings to become a giant of his time, and he deserves better than that.

ACKNOWLEDGEMENTS

My thanks go to:

Mrs Frank Dawson (Fev) for her warm support.

Vin Callcut and Neil Clarke of Broseley Local History Society for their editorial input.

The History Press for being a very clever pleasure to work with.

Ironbridge Gorge Museum Trust, Cyfarthfa Museum and Art Gallery, and the Royal Society of Arts for illustrations.

I

BEGINNINGS

The father of John Wilkinson, Isaac Wilkinson, the first of this family of ironmasters, probably came to Cumberland from Washington, County Durham in the late seventeenth century, but there remains some uncertainty about his origins. Recent research by Janet Butler[1] indicates he was born in 1695, the youngest of six children of John Wilkinson and Margaret Thompson who were married in Washington on 27 June 1678. A bishop's transcript of a 1705 entry in the parish registers for Lorton, Cumberland records: 'Isaac, son of John Wilkinson, baptised 24 January'. However, this might refer to another Wilkinson family, for if one accepts Janet Butler's dates, Isaac would have been 10 years old at this time. The further evidence, of his stated age of 80 years at the time of his death in the Bristol Register of Burials for 1784, must also be considered.

At an early stage in his adult life Isaac was a known dissenter; if he grew up with these beliefs in a family of dissenters, baptism in the established church would not have been possible. On the other hand, it may be that he developed these ideas later and that his parents, at the time of his birth, were conforming members of the Church of England.

It has been suggested, because of his subsequent close relationship with the Quaker William Rawlinson of the Backbarrow Company in south Cumberland (whose father had documented links both with the Bristol merchants and the Darbys of Coalbrookdale),[2] that Isaac came north to Cumberland from the Midlands and developed his religious views from an earlier beginning. He certainly moved south to Wrexham in his middle life, but whether that was a return, or another beginning, remains uncertain. We do know that he died in Bristol in 1784 but, in the meantime, there is further evidence for his northern roots.

Firstly, we know Wilkinson is a northern, rather than a Midlands, name. The church registers in the Lake District of present-day Cumbria are full of Wilkinsons, and the IG (Mormon) Index for the old county of Cumberland lists literally hundreds of them. Secondly, there is documentary evidence to show that Isaac came to the Backbarrow Company, on the River Leven between the southern end of Lake Windermere and the sea, from Little Clifton in Cumberland in 1735.[3] Little Clifton is in Workington parish, some 3 miles due east of Workington town, and about 8 miles by road north-west from Lorton. It lies in the mouth of the broad vale of the northward-flowing River Marron, a couple of miles from its confluence with the Derwent.

There is a church record, however, from the parish of Skelton, also close to Workington, for the year 1727 which records: 'January 20th: John, son of Isaac Wilkinson and Ann his wife, baptised.' If this is our Isaac, he was married by the age of 32 (or by 22 if one accepts the alternative evidence) to someone called Ann, whose maiden name is unknown. There is further confusing evidence for that year, too. From Brigham church, a village in the area just to the west of Cockermouth, comes a record that indicates Isaac was married there on 9 September 1727 to Mary Johnston, by banns.[4] He was 23 years old at the time. The date and the name of his wife, but not his age, agree with Janet Butler's evidence. If both records are accepted for Isaac then two things follow. First, the baptism record would mean that Isaac's dissenting ideas could not have developed fully by that time, since his child was baptised into the Church of England. Second, his wife Ann of the January record had died, possibly in childbirth, before he married Mary in the September. A possible explanation for some of the confusion begins to emerge.

Isaac's first marriage, sometime before January 1727, the date of the baptism of the John above, is to a woman called Ann about who little is so far known. She dies in childbirth and the infant John (who may or may not have survived) is baptised. It is perhaps this tragedy which turns Isaac away from the beliefs and practices of the established church. As a young widower, he meets Mary Johnston and marries her later that year. The following year their first child is born, but there is no church record of this birth or baptism because the father is now a dissenter. Such a scenario would be supported by all the evidence quoted above, with the one discrepancy of Isaac's age.

It is worth repeating here the folk memory still circulating in the Workington area of the birth of John Wilkinson: in a cart, when his mother was returning to her home in Little Clifton from Workington market where she regularly went to sell her farm produce. The birth, in such circumstances, was of sufficient notice to register the local view that someday the baby 'wod be a girt man'. Such stories handed down by word of mouth are surprisingly enduring, often rooted in fact though embellished in the telling, and stand more as an indicator than as evidence.

This story is sometimes used to support the idea that Isaac's wife (or second wife) was a strong and healthy woman, which is likely to be so since she went on to bear him five more children. It also supports the tradition that the Wilkinson family roots were in farming, even though in his early 30s Isaac was being described as an iron founder. Little Clifton, too, is in the middle of that favoured livestock farmland between the Cumbrian Mountains and the Irish Sea, where the young sheep and cattle born on the fells and in the mountain valleys come to be fattened. As Ron Davies describes:

> The area of Little Clifton today is completely by-passed by newer and faster roads, so without the aid of a detailed map it is for a stranger virtually impossible to locate. It lies cheek-by-jowl to Bridgefoot village which is set upon the River Marron, a pretty spot, boasting a secluded and ancient water-powered iron forge with an attendant weir and mill house.
>
> The old furnace where Isaac worked stood about half a mile south from Little Clifton, but today there are no outward visible signs of such, though cinder is seen in fairly large quantities and finding a lump or two of iron is no problem.
>
> As one would expect, the site is known as Cinder Banks, a name which has been adopted to a new bungalow recently erected upon the site. Across the field to the west of Cinder Banks is Furnace House. It now stands empty in a long and lonely lane and was probably used in days past by managers of the ironworks and possibly the Wilkinsons ...[5]

From Furnace House, the ground slopes down gently to the River Marron and its old mill half a mile away. The view beyond to the east is across gently rolling country, the low ridge separating the Marron and the Lorton vales in the foreground and the rugged peaks and ridges of the high fells of the Lake District on the skyline beyond. Still countryside of small farms, it will have changed little since Isaac's time.

There are eighteenth-century records of an iron furnace at Little Clifton, and it is likely that Isaac learned his iron-making skills there while his wife ran a small farm or holding that was their home. The Workington church registers from the 1730s record the christenings of several children of a 'certain Charles Reeves of Clifton Furnace', suggesting the place was a well-known and important focus in the area at that time.

Isaac is first described as an iron founder in an agreement, signed on 25 July 1735, between the Backbarrow Company (an established iron-making business in what was then known as Lancashire-over-the-Sands) and 'Isaac Wilkinson of Clifton in the County of Cumberland, Founder'.[6] It is a fascinating document, and makes clear immediately that the Backbarrow Company were contracting with an experienced and established craftsman. He undertakes:

> ... to cast for them all kinds of Cast Iron Ware whatsoever and what Quantities thereof as they may require him to cast at Backbarrow and Leighton Furnaces

for the Term of Twenty One Years (and it shall not be Lawfull for him to leave the said Business during the said Term upon any account if they think fit to continue the same) at the following Rates being sound and merchantable goods viz Pots and Pans of all sizes at Two Pounds Seven Shillings and Sixpence p Tun Girdles and Boshes at One Pound Four Shillings p Tun Backs Grates and Heaters at One Pound p Tun Weights at Fifteen Shillings p Tun Waggon Wheels at One Pound Eighteen Shillings p Tun and any other kinds of Work at Proportionable rates, the said Isaac Wilkinson finding all kinds of Tools Utensills and necessaries whatsoever requisite for Casting the said Wares at his own proper Costs and Charges, the said John Maychell William Rawlinson and James Maychell finding a Casting House of Twenty Yards long and Ten Yards wide for the said purpose …

Casting house fronting a blast furnace; a familiar sight for John Wilkinson.

Isaac, then, did not learn his iron-founding skills at Backbarrow, nor did he come here as a youngster to learn his trade. He is, at this point, 40 years old with enough experience at the Clifton furnace to give him an impressive range of casting skills. Moreover, he has sufficient standing to negotiate a compensation clause in his contract should it be terminated, and from the beginning he is pushing his employers towards innovations. The contract continues:

> ... But in Case the said John Maychell William Rawlinson & James Maychell do find the said Business not beneficial to them then it may be Lawfull for them at any time to make void this agreement provided they employ no other Workmen afterward in the same way and do pay the said Isaac Wilkinson Fifty Poundes for full Damage and Satisfaction in procuring Toolles; And it is moreover agreed that if the said John Maychell William Rawlinson and James Maychell do incline to have the abovesaid Wares made by an Air Furnace in the Intervalls when their Blast Furnaces are out the said Isaac Wilkinson hereby covenants to build the same at his own Charge and to cast the Wares at the abovesaid rates but not to find the Fuel for that purpose ...

There is another folk memory, told in the Backbarrow area, of Isaac Wilkinson being paid in part by his employers in molten metal to be used for his own purposes, and of him carrying it in pots from the furnaces across the road to moulds at his house. This has tended to be dismissed by commentators who understand the quick-cooling loss of fluidity of molten iron. Such memories become more feasible, however, in the context of this early reference to an air furnace, in which the metal could be reheated and further refined before being poured.

Information about Isaac's subsequent work at Backbarrow comes from the account books and journals of the Backbarrow Company. They show that his early energy and drive are impressive; building of the casting house, the 'new pothouse', begins in December 1735 and continues through the winter.[7] There is a payment against it of £45 11s 10d in February 1736, and Isaac begins casting in July even though the roof is not finally slated until September. His first quarter's wages are paid the same month and a new account for 'Isaac Wilkinson Potfounder' is opened, which shows a production, by the following February, of some 60 tons of pots, pans, backs, girdles, plates and wheels.

Later that year, he proposes another innovation to his employers. He has identified a marketing opportunity for an improved type of box smoothing iron, is confident of his skill to manufacture the new product himself, is keenly aware of the competitors in the market and what must be done to outmanoeuvre them, is clear about how the release of his new irons onto the market should be controlled and of what the price should be. His written proposals are accepted with only minor modifications and signed by all parties in an agreement dated 18 October 1737.[8]

This document identifies Isaac Wilkinson as a skilled iron founder certainly, but also as a potential entrepreneur with imagination and business flair; qualities that, from this point on, recur throughout his life. It also outlines the unusual relationship he was able to establish with his employers, the Backbarrow Company: the company is producing iron which they sell as Bar Iron by the ton, or make into iron products ('Cast Iron Wares'). Isaac is the skilled founder employed by them to make the cast-iron wares, for which he receives wages. But he is also allowed to sell for his own profit a proportion of the cast-iron wares he has made, under an arrangement whereby he buys back from the company for resale his own products at an agreed rate per ton of wares, the rate varying with the type of product.

For his improved box irons, for example, the rate he proposed was £12 or £13 per ton of wares, raised to £14 a ton in the agreement. He also asked his employers for sole rights for the sale of these box irons, which, since he took out a patent for that product the following year, it seems likely he was granted.[9] It is the manner of the moulding of the one-piece box that makes his smoothing irons innovatory, and the fact that they can be made from a 'melted fluid' of 'any mixt metal' indicates a further use for the air furnace in which any old metal could be re-melted.[10] The wording of the patent is revealing:

> ... my said metallick boxes, both bottom, top, sides, and the barrs within them, consist of one entire piece of any cast metall, either iron, brass, copper, bell metall, or any mixed metall, and are made and performed from a melted fluid of any of the said metalls cast into a mould invented for that purpose, and then ground and finished in the same manner as other box irons now in use.

Two interesting questions emerge at this point. First, to what extent was an iron founder, in those early days, looking towards the domestic market for his profits and as an outlet for his products? How far is his imagination and inventiveness focussed on the domestic scene? The list of Isaac Wilkinson's cast-iron wares suggests that the domestic market was important. Pots, pans, fire backs and grates, weights and smoothing irons are listed among his products. Second, what role did his wife play in engaging his attention on the need for an improvement to the box iron then in use? Its old construction of separate plates bolted together could allow hot ash or small cinders to drop out onto the ironing. Perhaps Isaac had some personal experience of this.

From earliest times, forges and furnaces were blown by leather bellows; the smaller ones hand operated, the larger ones, as at Backbarrow, attached to a cam wheel driven by a waterwheel. Servicing and replacing the leather airbags, which became creased and worn from constant use, was a considerable recurring cost. The Backbarrow Company journals show payments for 'tanned hides for bellows' in December 1736 and April 1737, soon after Isaac Wilkinson arrived there.[11] In the autumn of 1737, however, he changed, indefinitely, the dependence of his employers on leather bellows, in a step that at once demonstrated his imaginative flair and his iron-making skill.

The journal of the Backbarrow Company for 1737 has the following detailed entry: 'Backbarrow Forge Dr to Acc/t of Cast Iron Wares the sum of £6- for a pair of Cylindrical Cast Iron Bellows, put up in Septemr 1737 being computed at ½ a tun and valued at £12 per Tun ... £6-.' There is evidence that the company was enthusiastic about this innovation, was prepared to support it financially and wished to celebrate its arrival, too. There are account entries, round this date, for fourteen days' day-labourer payments at 1s a day 'to George Walters about Iron Bellows etc', and several transfers from one account to another of iron 'for new Iron Bellows'. Particularly interesting, and showing beyond doubt the enthusiasm of the company for this improvement, is an entry in 'sundry disbursements' for September 1737: 'For Ale ordered by the Masters on occasion of the Iron Bellows £3-.'[12] It was obviously a signal event.

There are two further significant records. Firstly, on 1 October 1737: 'By forge, for iron used about Geering the new Iron Bellows 1c. 7st. 12lbs ... £1.11.9d.'[13] Secondly, on 27 December the same year, when the forge was also charged 'for a pr of cylindrical Bellows & Appurtenances' weighing '18 cwt'.[14] The former of these entries could relate to repairs or improvements to the first iron bellows installed, but it seems probable that the latter refers to a second pair of bellows at another hearth. Overall, it is clear that Isaac Wilkinson was using iron bellows, designed, manufactured and installed by himself, for forge and foundry work at Backbarrow in 1737 – some twenty years before they came into use elsewhere.

In this context, too, the second part of his 'Box Iron Patent' of 1738, which is puzzling and often ignored, begins to make sense. For he includes in it another item, which is difficult to relate to box irons, and is described as his 'Bellows of Cast Metal for Forges, Furnaces or any other works ...'[15] His so-called 'Iron Bellows Patent' did not appear until 1757 – the date generally credited as marking the introduction of iron bellows into the iron-making industry – but that patent application is a carefully elaborated description for a 'machine or bellows to be wrought by water or Fire Engines ...'[16] By that time, of course, he had two decades of practical experience of iron bellows behind him.

It is surprising that, in those intervening years, no one stepped in to steal the invention, particularly in view of the enthusiasm with which the Backbarrow Company initially adopted it, which one would expect to lead to a wider broadcasting of its use. It may be that, in spite of William Rawlinson's contact with the Midlands iron-making world, this far outpost of Lancashire-over-the-Sands, difficult to access by road, though less so by sea, retained its isolation and integrity until the 1770s. By then, Isaac's iron-making sons John and William were living and working in the Midlands whilst retaining strong links with their roots.

During these early Backbarrow years, Isaac's family had not grown beyond the two boys born in Little Clifton in 1728 and 1730 respectively, John and Henry. It is as though he needed to pour all his energy into the early opportunities this new work provided. Things were about to change, however. In 1741 a daughter, Mary, was born; in 1744 a third son, William; and two further daughters, Sarah, in 1745, and Margaret, whose birth date is uncertain.

Tradition has it that Isaac lived throughout the Backbarrow years at Bare Syke, a substantial family house, gardens and orchards belonging to the Maychells, across the road and a little south of the furnace; close enough to supervise the work continuously but upwind of the furnace's fumes. There is reference in Backbarrow documents, long after Isaac was dead, to 'Wilkinson's House', which is likely to be Bare Syke. Ask after Wilkinson's House of any old resident in Backbarrow today, and they will point you to it. The impact of the family on the place has been enduring.

Isaac's two older boys, John and Henry, grew up at Backbarrow – youngsters of 7 and 5 when they arrived, young men of 20 and 18 when they left – and throughout their early adolescence they had no other siblings in the family to consider. It was a marvellous place for two such lads, quite apart from the fascination of the furnace and forge which Isaac would certainly involve them in as they grew older. Backbarrow lies in a gorge section of the River Leven, about a mile downstream of the point where it empties out of the foot of Lake Windermere. To this day, it is a clean river of waterfalls and pools where salmon wait, with extensive oak woods, rich in wildlife, spreading into the valley on both sides. Throughout his life, John particularly, and in stark contrast to his preoccupation with the noise, heat and smoke of his iron-making, responded to wilderness and water with a spirit that anticipated the later Romantics; he created his own corner of paradise in later years at Castle Head, not far from Backbarrow. Perhaps this is where it began, where he and Henry were free to roam, where this wildness and wet created a pattern, an ideal, for his later life.

Isaac and John at Backbarrow – an imagined picture

'Father, Father, the gentlemen are coming, the gentlemen are coming!'

John hurtled up the cinder track towards the furnace buildings as fast as his legs would carry him. He had been sent down to the house for bread and beer and had seen the two horsemen round the bend on the riverside road, half a mile downstream.

His voice, not yet quite broken, was shrill enough for Isaac to hear above the loud breathing and groaning of the furnace, and the creaking of the wheel. He came to the wide, doorless entrance of the casting shed with a twinkle in his eye and caught his son round the shoulders as he arrived breathless.

'Whoa lad, steady,' he said, kindly enough. 'What gentlemen?'

'It's Mr Rawlinson on the black stallion, and Mr Machell, I think …'

'Here, put this on,' said Isaac, handing him a heavy leather apron similar to his own. 'And put them clogs on. We're going to run iron. Them'll wait.'

Father and son moved back from the sunlit doorway into the inner darkness of the shed and picked up their long-handled ladles. Five other men stood by, waiting. Isaac looked round quickly, saw that the moulds for utensils and implements they had been preparing with fine sand for the last few days were all neat and ready, noted that the sand bellies and channels were clean and smooth, and gave the signal to the man standing by the furnace with a very long probe to begin.

The man pulled a darkened glass shade down in front of his face and reached forward with his probe for the plug at the bottom of the furnace. There was a moment's pause before he found it. Then, with a smooth movement, he withdrew the plug and stepped backwards. A thick, bright orange stream of molten iron came curving from the furnace through the channel and into the first sand belly, filling it quickly and overspilling into the narrow channels and the smaller bellies around it. John, standing beside his father and the other men, watched in wonder as the living metal slowed and became still, then changed colour as it began to cool. It drew out at his feet the image of a pot-bellied sow with her piglets feeding from her all around. It was a new creation that always thrilled him.

The man with the long probe now diverted the molten flow, to create another sow and piglets in a second set of sand channels and bellies alongside the first. When this was almost complete, he glanced at Isaac for the signal before diverting the bright metal into a large cauldron standing close by. As it began to fill, Isaac stepped forward, dipped his ladle into the cauldron and carefully carried the molten metal the few strides across the floor to pour it into the box moulds. As his ladle emptied, John came to stand beside him to continue the process.

Timing was everything to ensure a continuous flow of bright hot metal into the moulds. The orange flow from the furnace had begun to falter, and at another signal from Isaac the man with the long probe carefully placed a new bung directly into the furnace tap. As the flow ceased it became darker in the shed, the metal in the moulds now a dull red. In the bright sunlight of the casting shed doorway John saw the two gentlemen standing with the light behind them, smiling as Isaac moved towards them, taking off his apron. Then John was sprinting back to Bare Syke for the bread and beer he'd been sent for and forgotten.

Henry is a shadowy figure and virtually nothing is recorded about him. It has been suggested that he was born impaired, handicapped in some way, but this has to be speculative. He did not follow his older brother to school, despite the fact Isaac was careful and determined in the education of his children; there is no record of a marriage; and he remained within the embrace of his father's house until he died at the age of 26.

There is one other haunting piece of evidence. Carved into a vertical face of the hard slate behind Bare Syke, and still visible in a place where Isaac is said to have espaliered his fruit trees, are the initials 'HW 1745' in a fine elaborated script – obviously the handiwork of a competent scribe. If Henry did this he had certainly learned to write and might have been taught within the family. Or did John inscribe it for him? What could be John's initials 'JW' (possibly 'IW') are cut into the same face of rock close by.

It is interesting to speculate that, had Henry's life been impaired from birth, might this not be reason enough for the eleven-year gap before the arrival of the next sibling? The more so if there had been earlier birthing tragedies. Perhaps Isaac and Mary had decided to have no more children, even before they came to Backbarrow. If so what changed that? Or was their daughter Mary a happy accident which carried them into the sunlight again? Such things would have implications for the happiness of the family and for the atmosphere in which the children grew, particularly for John who was alone with Henry for so long.

The possibility, of course, remains that John could have been the surviving son of Isaac's deceased first wife. If his first wife, Ann, had died when John was born, and Henry was the damaged firstborn of his second wife, Mary, Isaac might have decided to have no more children. All his considerable energy would go into his iron-making. Then his daughter Mary was born and everything changed again.

In his old age, when he was living in Bristol, Isaac signed an agreement with Thomas Guest of Dowlais Furnace at Merthyr Tydfil and Thomas Whitehouse Ironmonger of the city of Bristol, for the casting and manufacture of iron goods which contains the following clause:

> ... it is also hereby agreed ... Thos Guest and Thos Whitehouse ... to pay the said Isaac Wilkinson <u>and Mary his wife</u> during their natural lives one shilling per ton for every ton of Iron ... made at the said works from coakes over and above 15 tons per week on average through the Blast ...[17]

The underlining is mine. It is not, of course, proof that a first wife died in those early years. It does establish that in his old age Isaac had a wife called Mary, and, if the Workington church record identifies the same Isaac, then the Ann named there as his wife must have been dead. It supports the possibility that John's natural mother might have died when he was very young, that he was brought up by a stepmother, that William – with whom he quarrelled bitterly and irreconcilably in his later life – was his stepbrother, and that his closeness to Henry might simply have been a matter of their closeness in age.

In those later Backbarrow years, John left Henry to go to school. Though at precisely what age is uncertain, and since Isaac sent him to a Non-conformist school at Kendal, it is likely Isaac and the family were known dissenters by this time. The church schools, and there were few others then, simply would not be available to them, so they were fortunate in having a good school run by a remarkable man close by.

The Reverend Caleb Rotherham was associated with the Unitarian Academy at Kendal from its inception in 1733, until his death in 1752. On 27 May 1743 he was admitted Master of Arts of the University of Edinburgh, followed immediately by a DD (Doctor of Divinity), awarded on public disputation 'On the Evidences of the Christian Religion'.[18] His academy, however, was not a theological academy, and two-thirds of his students were never intended for the Ministry. It was a place for young men, as opposed to boys. He charged 8 guineas a year for lodging and board, and 4 guineas for learning. The young men were required to provide their own fire and candles and wash their own linen: 'They go through a whole course of Mathematics ... I have a distinct consideration for that branch of instruction ...'[19]

Outside the universities of the time, which, as with Church establishments, would be closed to him, it was probably the best education available to Isaac's sons. Because of the building and engineering demands inherent in the iron-making processes, Isaac would recognise the value of a good grounding in mathematics. There is plenty of evidence in John's later life that it served him well, and that his handwriting and use of language had also benefited.

He, though never Henry, is in a list of Caleb Rotherham's students, which has been dated at *c.* 1745 as a consequence of other names included in it. The precise dates of John's education at the academy are not known, though since he would be 18 years old in 1745, that list was likely to have been drawn up towards the end of his time there. It is recorded that Isaac later sent his son William to school at the Warrington Academy when he was 14 years old, so it is reasonable to suppose John started at the same age and spent four years at Kendal.

Two other young men at the Kendal academy about this time were James and Robert Nicholson, sons of Liverpool merchant Matthew Nicholson, who was himself cousin to Edward Blackstone, one of the original founders of the Unitarian Chapel in Kendal. His widow, Ann, married Caleb Rotherham in

1746, after the death of his first wife. Matthew Nicholson, with his Kendal links, may therefore be the 'respectable merchant' to whom John was said to be apprenticed on his return from school, 'and with him continued about five years …'[20]

In 1740, when he would have been considering the education of his eldest son, Isaac was spending more time at Leighton furnace, also owned by the Backbarrow Company, where he began casting and trying guns. How far this was driven by an urge to continually push out the horizons of his work, or by his employers' belief that there may be good profits in it, is difficult to assess. There is, however, evidence of restlessness in Isaac's life at this time, and also the first signs of differences of opinion with the company.

On 2 February 1741, shortly before the birth of his daughter Mary, Isaac signed a lease with James Machell for a run-down corn mill and kiln on the east side of Backbarrow Bridge, and a dwelling house with outhouses, orchards and gardens on the west side.[21] Bare Syke is on the west side and the description of the property fits, though there is a discrepancy in precise location. Does this mean that Isaac, up until this point, had held his family house by grace and favour until this lease gave him occupancy as a tenant? And if so, was it a move by the astute Isaac towards greater security? Or is the dwelling house in the lease another property, which, since he was living at Bare Syke, he subsequently sub-let? Certainly by 1753, when Isaac sought to terminate this lease, one 'Widow Taylor' was living there, but by that time he had been gone from Backbarrow for five years.

The corn mill was fitted out with new grindstones, suggesting he used it to finish the cast-iron wares like box smoothing irons, which he sold on his own behalf, providing him, at the same time, with his own commercial premises. If disagreements were beginning to emerge with his employers, this would be important.

The casting of guns was not going well. The accounts show occasional, rather than regular, evidence of the purchase of gunpowder 'for trying of cast guns',[22] some of which burst on proving.[23] There is also an increasing tension between William Rawlinson and the other partners at this time. William Rawlinson had been Isaac's chief support in the Backbarrow Company, certainly for the widespread sale and distribution of his cast-iron wares, and may have been responsible for the ill-starred sortie into gun manufacture. By 1743 the partners had had enough. In a new agreement with Isaac, dated 14 March, there are indications that a substantial conflict is being resolved:

Let it be remember'd that Backbarrow Company having sundry articles under Consideration relating to Isaac Wilkinson … and have concluded &

agreed with him as followeth, viz That Damage which the Company have suffered by Casting of Guns, shall be Ballanced by the workmanship of Casting Hammers & Anvills at Leighton last Blast, so that all Demends on both sides in these respects are to cease and be evened ...[24]

'Let it be remember'd' has an admonitory tone, which is significant, and it would be fascinating to know what these 'sundry articles under consideration' were and whether William Rawlinson stood beside Isaac at this time. The future involvement of each of them in the Backbarrow Company was to be short-lived. A statement of William Rawlinson's alleged debts to the company, amounting to £3,369, was drawn up by the manager in February 1747. A counter-claim by Rawlinson demanded the production of the accounts and the sale of stock to meet the company's debts to him, and the dispute was only settled when the Machells undertook to buy out William Rawlinson's moiety in the company in an agreement dated 8 April 1749.[25]

By this time, Isaac Wilkinson had gone. Following the new agreement of March 1743, the journals continue to show an Isaac Wilkinson account with the company each year until 1748, but in 1747 a major dispute erupts. By now, Isaac is in partnership with William Rawlinson's brother, Job, and two other men in a new iron-making venture, the Lowood Company, about 1 mile down-river from the Backbarrow works and potentially a serious competitor to it. Isaac was still bound to the Backbarrow Company by his 1735 contract, and they went so far as to obtain lawyers' opinion as to whether he was still in their employ

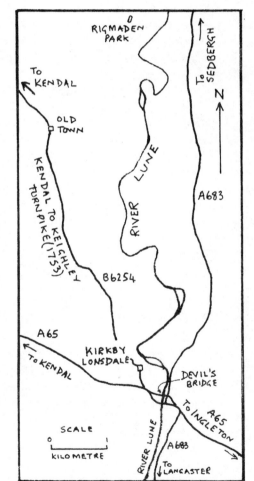

Map showing the location of Rigmaden.

and could legally do this. They were advised that, provided he continued to work for them, there were no grounds for legal action against him, but the end was in sight. His account with the company was drawn up and balanced to 25 March 1748; after twelve years of continuous business his name appears no more in the registers of the Backbarrow Company.

The terms under which he was released from the 1735 contract ahead of the twenty-one years he was required to serve, are not known. Fell, who obviously had access to primary sources now lost,[26] tells us that Isaac and the other original partners in the Lowood Company disposed of their interest to two local men in 1749, and there is some later evidence to support this.[27] The Backbarrow years, consequently, came to an end in 1748, at which point Isaac Wilkinson and his family moved some 5 miles south-east to Wilson House, near Castle Head Hill beside the River Winster.

2

WILSON HOUSE

By the time Isaac moved to Wilson House, near Lindale at the mouth of the Winster Valley, he was 44 years old and his family was complete. John and Henry were young men by this time, Mary was 7 and William 4, and the two youngest girls, Sarah and Margaret, were still toddlers. John left the family home soon after he left school and would quickly complete his apprenticeship with a Liverpool merchant.[1] There are no records of the details of the apprenticeship, but Isaac might have timed his move from Backbarrow in anticipation of his eldest son's return. Did he by now have a vision of a family business focussed on iron-making?

Both Stockdale, in 1872,[2] and Dickinson, in 1914,[3] say that Isaac moved to Wilson House because of the abundance of peat in the vicinity. He had certainly experimented with peat in various forms as a supplementary fuel at the Leighton and Lowood furnaces, and perhaps saw it as an easily available supplement to charcoal – supplies of which, even in the well-wooded Furness area of Cumberland, had by then become difficult to obtain. It is interesting, in this context, that when John later in his life returned to Wilson House, in 1778 and 1779, by then a wealthy and successful Ironmaster, he too experimented with peat as a furnace fuel – a detailed account of which can be found in his letters to James Watt at that time.[4]

Stockdale is quite explicit about Isaac's experiments with peat at Wilson House, saying he was unsuccessful and had to revert to smelting with charcoal, although he says that a mixture of well-dried peat bricks and charcoal was being used at the Backbarrow furnace at the time he was writing (1872).[5] Isaac certainly operated a furnace at Wilson House, as the following agreement makes clear:

Cartmel, October 30th, 1750. – Be it remembered that this day Robert Bare of Cartmel Church Town has sold to Isaac Wilkinson, of Wilson House, two hundred tons of wett flatt iron ore, to be put on board at Lousay, the said Isaac Wilkinson promising to pay for the same twelve shillings for each ton, but in case the said Isaac Wilkinson does not approve of the said ore, that then he is only to have fifty tons of the said ore, he giving the said Robert Bare notice in April next, that he will have no more than the said fifty tons. If no notice is given then, he to have the whole two hundred tons, the said Isaac Wilkinson

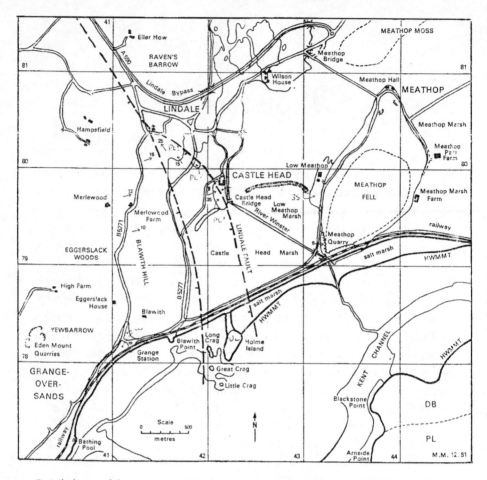

Detailed map of the area around Castle Head, the estate of John Wilkinson.

paying for the same on the second day of February, 1751. Signed, Robert Bare, Witness – Walter Cowperthwaite. Isaac Wilkinson.[6]

Stockdale in 1872, like Fell in 1908, had access to primary documents which have since been lost, and he describes the signatures in this agreement as being 'in excellent mercantile hands'; evidence that he held the original. But for all that, there has to be some doubt that he quoted it correctly. Isaac is required to pay for the whole consignment before the expiry of the notice period, which allows him to buy only a quarter of it if he doesn't like the quality. This is a contradiction in terms and likely to be an omission or oversight by Stockdale rather than Isaac. The let-out clause is typical of the business shrewdness evident elsewhere in Isaac's dealings, but the agreement has another interesting focus.

The ore is coming in by water, probably by coastal barge or flatte, and Isaac is committed to the first load whatever the quality. At that time, the River Winster was open to the sea and tidal the 2 miles up to Wilson House and beyond, giving Isaac access to cheaper water transport for his raw materials in, and his cast-iron wares out, an important consideration in view of the poor roads into the area, and a further demonstration of his business acumen.

It would be at this time that John first discovered Castle Head Hill, a wooded rocky knoll round which the Winster curled, standing a mile downriver of Wilson House en route to the sea. It remains, to this day, a striking feature in the landscape; a faulted block of limestone standing up in sharp contrast to the flat valley floor sediments all round. At that time, the spring tides had unobstructed access into the sea end of the valley, and twice a month they flooded all around the hill leaving it as an island at high water. Southward into the sun, the shining mud and sand of Morecambe Bay stretched away to Lancaster. From there the Guide-to-the-Sands led, at low tide, long strings of people and carts safely across the channels and through the treacherous sinking sands of the bay to the Cartmel and Furness coasts. His is an ancient office, a Crown appointment dating back to medieval monastic days to provide a quicker and easier route between Lancashire and the Furness region of Cumberland, and it survives still.

The hill of Castle Head stands sentinel at this point on the coast to further access into up-country Westmoreland and Cumberland, and with dramatic skylines behind it has a powerful presence. There is evidence of its occupation back through Roman times into prehistory both as a look-out point and a sanctuary, and John, here in his early 20s, came powerfully under its spell. He was to return to this place again and again throughout his life, building his house and, finally, being buried here.

John and Henry at Castle Head – an imagined picture

The two boys edged the raft into the middle of the channel where the water was deeper, and began to drift downstream. It was a homemade craft and not very stable. This was its maiden voyage. They had built it secretly the day before – four wooden spars tightly roped into a rectangle, some thin planks tied across to form a deck and a bundle of straw, wrapped in old bellows hide, at each corner for buoyancy.

They were big lads but awkward; in that stage between puberty and manhood where they were keen for an adventure. It was seven o'clock and birdsong was petering out as the sun climbed steadily on the left. They were round the bend below the farm now, moving sedately with the rock looming larger. It was called

Castle Head. They had wondered why. Henry had come up with the idea of ruins and the exploration was confirmed from that moment.

Henry turned round suddenly; the raft lurched and dipped deeply at the corner.

'Henry, Henry, steady. Sit still,' John shouted, trying to read the signs in Henry's face. They were good. His eyes were shining. He was happy. John relaxed a bit. Then, 'No, no man! Don't stand up, you'll tip us in!'

Henry had pulled his legs out of the water at the front and was on his knees on the planks, pointing with one hand to the rock ahead and trying to say something. John realised he was saying 'wall' over and over again, but was too concerned with balancing the raft to be able to look.

'Please Henry. PLEASE sit down again!'

'Wall, wall, wall,' continued Henry, but he had by now caught the urgency of John's concern and sat again with his legs in the water, still pointing to the rock with stabbing movements of his finger. They moved quickly in the current through the arch of the Lindale Bridge, into the old harbour pool where the banks widened and the water spread out and ran deep. John breathed a sigh of relief as they passed on without mishap, and looked up the steep side of the rock to where Henry was pointing. Sure enough, high near the top where the slope eased, almost hidden by brambles and woody scrub was the tumbledown end of a wall.

The wooded rock was now towering above them, and John looked for a place to beach the raft. He signalled to Henry that the best thing was for both of them to get into the water and push the raft to the bank. With an exuberant yell, Henry leapt off the front of the raft, which then shot up in the air, throwing John backwards into the water. Dripping wet and laughing like fools, they dragged the raft up the bank.

It was dark in the wood and quite cold in their wet shirts. They took them off and tied them around their waists by the sleeves and immediately felt warmer. Henry spun round and his wet shirt flared out like a skirt. John grinned and led him into the wood. They found a deer track that ran upward through the trees and they climbed in silence. With terrible suddenness in that quiet place, a loud grunting bark startled them from above, but immediately they knew it was a roebuck challenging this intrusion into his territory. Henry laughed happily and began to pronk and grunt in return.

They came out of the trees through a thicket onto a broad rocky step. Before them the shining levels of the sands stretched away to a misty coastline and to ascending horizontals of blue-grey ridges merging into the Lancashire sky. The place seemed untouched by man, primeval, awesome. John looked, holding his breath; Henry also silent, also looking, knowing something he could not articulate but which John could sense, nevertheless. He laid his arm across Henry's shoulder and felt the tightness ease.

In that moment, John knew that he would come back here, that in some way he couldn't understand he was a part of this place, and that Henry had sensed it too.

There are questions as to precisely how Isaac eventually blew his furnace and forge at Wilson House. Although the site is close to the river, the gradient there is minimal and the river meanders with little current across a flat valley floor. Just upstream from Wilson House, however, there is one slight break in slope, towards which the remains of a wide ditch can still be seen cutting through a peaty field. It might just be possible to obtain a sufficient head of water, by way of the ditch, to turn a cam wheel to power a pair of iron bellows at the furnace adjacent – which then must question Stockdale's account of the same ditch, which he says was cut as a canal for a boat to carry peat to the site.[7] An iron boat at that, the first iron boat ever built, and which, at this date, would have to be credited to Isaac rather than John. John certainly built an iron boat some thirty-five years later on the Severn and could have used one at Wilson House after that, but Stockdale's account of an iron boat here at this time is unlikely.[8] He was, after all, gathering evidence for this assertion 120 years after the event.

Isaac certainly used water power close by in Lindale, where the valley floor sediments meet the faulted slates and limestone and the slopes rise steeply. According to Stockdale, he powered '... a large grindstone ... by a small waterwheel erected on Lindale Beck at a place called Skinner Hill, about a hundred and fifty yards above the higher public house at Lindale ...'[9] This is such a careful and detailed description that Stockdale must have had good evidence for it, which, unfortunately, he does not quote. But at precisely the location he identifies, the Lindale beck runs steeply downhill through a mini-gorge section, and in an adjacent field there is a flat bed of sediment behind an artificial bank; good evidence for the storage pond a waterwheel on such a small stream would need in drier weather.

In these Wilson House years, Isaac, though still principally engaged in the manufacture of iron goods, became once more interested in farming. There was little opportunity in the scope and size of his property at Backbarrow to do more than build a terrace and grow fruit trees, but his tenement at Wilson House extended to at least ten acres.[10] The needs of his agricultural neighbours along the coast and in the Winster, Cartmel and Lyth Valleys would surely have provided a market for the agricultural implements he began to make in his forge.

His second patent dates from this period, for '... cast metallick rolls for crushing, flattening, bruising or grinding malt, oats, beans or any kind of grain ...' It is dated 24 January 1753 and establishes that he was still at Wilson House at that time, since he gives Wilson House as his residence in the application. Another very interesting point emerges from the application. Isaac names himself as 'Isaac Wilkinson, gentleman'; not pot-founder, or iron-founder, but gentleman. That one word, used at this particular point in his life, might tell us more about his personality and ambitions than pages of lost letters. It tells us, too, that in the social code of the day he had at least the means to dress and to live in a certain style. To probe that piece of evidence further it is necessary to examine what was happening to his eldest son, John, about this time.

The only solid information is in the Kirkby Lonsdale parish registers. In 1755, at the age of 27 or 28, John married Ann Maudsley, the wealthy daughter of a landed family from Rigmaden Hall north of Kirkby Lonsdale in the Lune Valley. The register entry reads:

> No 17 Mr John Wilkinson merchant of Kirkby Lonsdale in the County of Westmoreland and Miss Ann Maudsley of Rigmaden in the said Parish of Kirkby Lonsdale, spinster, were married in this church by Licence from Richard Atkinson Clerk Surrogate this twelfth day of June 1755 by T Croft. This marriage was solemnised between us [signatures of John Wilkinson and Ann Maudsley] in presence of [signatures of Wilson Jn Robinson and Chrisr Wilson].[11]

Neither a Wilkinson nor a Maudsley witnessed the marriage. Were they absent because they disapproved? The fact that the marriage took place in Kirkby Lonsdale church is interesting, given John's dissenting background, and may be the reason why a licence was needed and why Isaac was not present. It could also suggest some persuasion on the part of Ann, or her family.

Opposition to the match from the Maudsleys could be expected, as a landed family who had held the Manor of Mansergh in which Rigmaden stands since 1661.[12] The Wilkinsons' origins in the recent past were as pot-founders. Isaac's description of himself as gentleman, and John's as merchant in the marriage record, begin to acquire new significance if the purpose is to woo and wed a Maudsley. If the Maudsleys had opposed the marriage, it might well have coloured John's attitude to the gentry in his later life.

There is another focus. Thomas Godsalve, Ann's ancestor who first purchased the Manor of Mansergh, was a Dutch merchant. He of course was long dead, but could there have been surviving sympathy in the Maudsley family, because of these roots, for a personable young man who was in love with their daughter even though he was in trade? Conversely, there might have been increased opposition to him as a reminder of a past they wished to hide. The evidence is with the former, and is found in another church register entry for Kirkby Lonsdale for 1756 when the child of their marriage, Mary, is baptised. The parents are recorded as 'of Rigmaden', which suggests they lived there after their marriage, which in turn suggests the support and sympathy of Ann's family – though the baptism of the child in church again smacks of compromise. The appearance of the Maudsleys in the Kirkby Lonsdale registers continuously for a hundred years prior to this date indicates their enduring and traditional association with the established church. Their acceptance of a dissenter into the family, and one with John's lowly background, would require a balancing consideration from him. How far good relations were established might be gauged from another record in the registers for the following year, when John is witness to the marriage of Ann's sister, Margaret, on 18 May 1756. It all points to a love match that was able to survive their differences.

Meantime, Isaac has left Wilson House. Whether John returned to Wilson House when he left Liverpool is not known, nor is it clear if his move to Kirkby Lonsdale was due to any difference of opinion with his father. There is an absence of information about him in the early 1750s, at a very important point in his life. He was a tall, strongly built young man with a good education and an apprenticeship completed. Though whether the apprenticeship was to some aspect of iron manufacture, or whether he served the iron trade simply as a merchant at Kirkby Lonsdale, remains unclear. How far Isaac's departure from Wilson House to the Midlands represented a setback to any vision he then had of a successful family iron business, or conversely, how far it provided a better opportunity to build it, is similarly unclear.

Isaac's friendship with William Rawlinson, his former employer at Backbarrow and a man with an established contact with the Darby iron-making family, perhaps endured and deepened in the Wilson House years. It is likely that Rawlinson advised him of the opportunities in the Midlands and of the availability of the lease on the Bersham blast furnace near Wrexham, a place with strong Darby connections. Isaac certainly took the lease from the Chirk Castle Estate in 1753, and the same year, as befitted a gentleman, he rented from Squire Yorke of Erdigg a substantial three-gabled residence in its own grounds called Plas Grono in the township of Esclusham Isaf.[13] The lease of the property remained with the Wilkinson family for the next twenty-one years.[14]

Wilson House, too, was retained and eventually became the home farm of John's Castle Head estate, though it was certainly Isaac's property in 1757 by which time he was resident at Plas Grono. Stockdale quotes an important document in his possession in 1870, listing encroachments onto common lands in the parish of Cartmel: 'Isaac Wilkinson, Bersham, for an encroachment at Wilson House, taken off the common to enlarge his fields, 10 perches at £21.0.0d per acre, £1.6.3d. Mem – Will pull down or pay before Easter.'[15]

John and his wife and daughter followed Isaac to Bersham in 1756, and rented a modest house in Wrexham Fechan 3 miles down the road. He is unlikely to have done so if he and his father had quarrelled; but any plan afoot at that moment between them to build an iron-making empire to rival the Darbys of Coalbrookdale has to be speculative. Isaac was certainly ambitious and very confident in his iron-making expertise, and John was a young man who had made a good marriage but with his way yet to make. The time was ripe. But then disaster struck.

3

BERSHAM – A NEW BEGINNING

Isaac had been living in Bersham with the rest of his family for almost three years by the time his eldest son, John, arrived with young wife Ann and baby daughter Mary, just a few months old. From what is known already about Isaac's drive and energy during the earlier years at Wilson House and Backbarrow, it is long enough for him to have established himself in the area and to have a forward plan. Glimpses of what that plan might have included emerge in the next few years and the likelihood is that it involved John.

There would need to have been a tempting prospect in store to persuade John to leave his merchant business in Kirkby Lonsdale and his wife's ancestral home at Rigmaden, for what amounted to an immediate step down in status in a new and untried environment, and the more so for Ann. The move from Rigmaden Hall to a small town house in Wrexham would not be undertaken by her without real hope of better things to come. But they were young and in love, and everything was possible; in that summer of 1756 none of them knew what pain and grief would soon overtake them.

John's brother Henry died first, on 26 June at Plas Grono, by then the family home, and was buried in the dissenters' graveyard in Wrexham. There are no records of a declining illness. He was 26 years old and unmarried. One can only speculate on the effect of the bonds here broken; the impact of this death on Isaac and his wife who had loved and protected Henry within the family fold all his short life; and on John who lost the close companion of his early days. The place of burial confirms Isaac's dissenting views at this date, and John later erected there a memorial to his brother. But fate had worse in store.

On 17 November John's wife Ann died, and he lost forever the radiance that for a few brief years had filled his life. She was 23 years old. His desolation is recorded on a memorial he placed on the wall of Wrexham parish church; an ornate white marble plaque inlaid with black. That it is there at all inside this magnificent church indicates at least a certain ambivalence in John's religious ideas at the time; an attitude which reappears in his later life in spite of his father's clear position as a dissenter.

Since Ann's death occurred within a few months of childbirth it is likely, despite the absence of documentary evidence, that the hazards associated with childbirth at that time – and the absence of anything but primitive medical techniques and treatment in the face of complications – were responsible for it. This might suggest that she undertook to move from Rigmaden to Wrexham when she was already a sick woman, though the physical rigours involved in the journey and upheaval so soon after her first baby could themselves have brought on a decline. Did John have to persuade her to move? If so, did he feel in part responsible for her death? What was his response now to the little baby who lived on as a constant reminder of what he had lost?

The evidence is that it was more than he could bear. Mary was put out to nurse and for the next few years was brought up in the family of 'Mr John Flint, Gentleman', who controlled the stamps and post in Shrewsbury, leaving her father to grieve alone and find a new purpose in his life. It is after this tragic loss that his focus as an ironmaster begins and it could have been the catalyst which drove him.

Soon after this, the paths of John and his father begin to diverge. It is as though John needed to establish his position and status independently of anything that might have come through Isaac – who by now was struggling at the Bersham furnace, with an unreliable water supply, a poor quality iron ore and a smelting process with coke which suffered from insufficient blast. It seems, too, that some of the underground pipework to and through the site was in poor condition, as witnessed an undated letter in Isaac's handwriting to Sir Richard Myddleton of Chirk Castle:

Sir I am informed it you have a firret at ye castle: which I should desire ye loan of it for a day or two: I will take partickler care of it and return it safely by ye bearer who is to bring it if you can spair it so long. It is only to put after a rabet thro' our pipes it walking throw in order to drive ye rabet thro' so as to discover if there be anything left in them: by this method we can tell where it is and cut ye pipes and take it out. I have aplyed for one to several places but cannot yet meet with any which makes me give you this trouble. If anything should befall it I will pay any prise for it ...[1]

Isaac was in his early 50s at this time. John is 28 and perceptive enough to see the limitations of working with a strong-minded father in an unprofitable business that needed substantial input and changes to turn it round. He decided instead to look for an opportunity to make iron in his own right, and at this point in his life had the confidence to move away from his father and to contemplate another future and a new life.

Part of that new life was a gradual broadening of his circle. It was about this time that he met for the first time Samuel More, a young apothecary based in

London, a year or two older than himself with a good classical education that boasted the addition of arithmetic and accounts. In the late 1750s More was gaining notice as an experimental chemist working with Dr William Lewis at the Society for the Encouragement of Arts, Manufactures and Commerce and, on 13 May 1761, was elected a member.

More and John, both unattached males, must have been established friends by then, for a few months later More introduced John to the society and nominated him for membership. It was More's first nomination and an important and prestigious step for John, who must have had a growing reputation for his iron-making innovations in order to be acknowledged by the society. It immediately gave him access, at this early point in his rise to power, to the elite of the manufacturing and commercial world, and to the many new inventions and forceful personalities that were a dynamic part of it. These, too, were the early years of a close and continuing friendship with More – greater than with any other of his contemporaries – which was to endure and deepen into old age.

Throughout this period, Isaac carried on making iron at Bersham and the evidence is that he still nourished the hope of establishing with his sons a new dynasty of iron-founders. On 12 March 1757 he filed an elaborate patent application for his iron bellows.[2] The wording included the following surprising description:

> … when full of air … the air is compressed by a pillar of water of a proper altitude … and forced out by the water through a pipe at any distance required, so that a furnace, forge, or any other works may be blowed from any waterfall or falls, or from a fire engine or engines to several miles distance … by means of a pipe being fixed to the machine, to force or convey the wind through to the said work …

The description of this device is significantly different from the 'Blowing Engine' John developed at his New Willey works a year or two later, which More described and a drawing of which is preserved in the British Museum (where iron bellows, regulating bellies and furnace are all in close proximity).[3] Isaac, it seems, was trying to convey the blast, once generated, considerable distances through underground pipes to the furnace. In this context his undated letter seeking to borrow a ferret from Richard Myddleton to check that his underground pipes were open, has new significance.

It is interesting that he took twenty years to develop this blowing machine from his simple iron bellows, and patenting it at this time suggests he had realised, since arriving in this hub of the iron-making world, that other ironmasters would adopt or modify it, as his son would do, and use it to their profit. He must have had confidence, too, in the survival and expansion of the Bersham furnace, provided he could improve the blast for coke smelting, for on 9 June 1757 he took out a forty-year lease from one John Hughes for all coal and iron lying under the nearby estate of Cae Glas, in Esclusham Uchaf near Llwyn Enion.[4]

The following year he sent his son William, then aged 14 but destined to become the third ironmaster in the family, to the school of another well-known dissenter, Joseph Priestley, some 20 miles away at Nantwich in Cheshire – evidence of his continuing religious dissent at the time and a decision that was to have important consequences for his family in the future.

Priestley was a scholarly and free-thinking young divine not yet ordained. He was 25 years old, unmarried, Yorkshire born, a product of Batley Grammar School and a graduate of the Daventry Dissenting Academy. He had already published a number of religious texts, which had attracted notice and had served two brief ministries at Needham Market in Suffolk, and in Sheffield. He suffered from a hereditary stammer, which must have made it difficult for him to preach a sermon and perhaps led to the kind of compensatory behaviour which made him 'too gay and airy'[5] for his Sheffield congregation:

> The pupils … thought of him first and foremost as an eccentric. Walking in a kind of disjointed, birdlike trot, Priestley chattered incessantly, stammering like a woodpecker. Even more disconcerting was the fact that the two sides of his face were so unlike as to cause a marked difference in his left and right profiles.[6]

His school at Nantwich flourished, however. It was established soon after his arrival there in September 1758 and Isaac's son, William, was to become one of the first pupils. Isaac was obviously party to information circulating among the dissenting fraternity, first because he knew of Priestley's school at a very early stage, possibly before it opened, and second because Priestley had an interest in mathematics and science in addition to his theological enquiries,

Joseph Priestley, innovative teacher and philosopher who received and reciprocated long-term support from John Wilkinson.

which would have been an important consideration for Isaac. He might also have liked the man he saw; an eccentric, certainly, but in the opinion of Josiah Wedgwood who knew Priestley personally, another rising star at this time and a man with an unmistakable touch of genius:

> Priestley spoke and moved rapidly; in private converse he was vivacious and fond of anecdote, often smiled, but seldom laughed; he would walk twenty miles before breakfast, carrying a long cane, and was a good horseman. He uses no action, no declamation, but his voice and manner are those of one friend speaking to another. In person [he] was slim but large-boned; his stature about five feet nine, and very erect ...

It seems he was also a strict schoolmaster, 'never giving a holiday on any consideration. His school and private tuition occupied him from seven in the morning till seven at night.'

William's 18-year-old elder sister, Mary, certainly liked what she saw. How many times in the next three years did Joseph Priestley walk or ride the 20 miles to Bersham on the excuse of offering William some private tuition?

Priestley left Nantwich on his appointment to the Tutorship of Languages and Belles-Lettres at the Warrington Academy in September 1761, and was ordained there on 18 May 1762. The move signalled the end of William's formal education and doubled the distance Priestley had to travel to see Mary. His salary at Warrington was £100 a year and came with a house, by which he was able to supplement his income by taking boarders, and Mary Wilkinson married him on 23 June that same year soon after he was ordained. She was 21 years old. He was 29.

Their first child, Sarah, was born the following year, and their next few years together at the Warrington Academy were happy, though towards the end of that time the first signs of a recurring problem with Mary's health appeared. Glimpses into their early marriage show Mary involved in the social recreations of that close academic world, obviously respected as a woman of 'sound culture and strong sense' – yet further evidence of the importance her father Isaac had attached to good education and independent thinking.[7] Although the wedding was at Wrexham her father did not give her away, an office performed by one of Priestley's pupils, and it is a strange omission in the light of other events in the Wilkinson family in 1762.

During these years, the paths of John and his father had further diverged. John's financial position would be based on what Ann Maudsley had brought to him as a dowry – which might not have been a large sum if there had been Maudsley objections to the marriage – plus any accumulated profits he had made from his business as a merchant at Kirkby Lonsdale. With the Bersham furnace barely profitable he could not expect, nor perhaps for other reasons have wished to ask for, financial support from his father. He would instead have been looking for potential partners prepared to back his drive and ideas with capital, and his need to start a profitable business quickly was urgent. It may be, too, that he chose to concentrate his search in the Coalbrookdale and Broseley areas, which were at the focus of the Midland iron-making developments with established water transport south down the Severn, as opposed to Wrexham and Bersham which looked north to Liverpool and the Dee ports. It also placed some 30 miles distance between his and his father's activities and that might have been important to him.

As early as 1756, John had been in discussion with Brooke Forester about the design of a new blast furnace on the Forester estates at New Willey, just across the river from Coalbrookdale and destined to become the first furnace of what would be their New Willey Company. John by this time would have made himself familiar with the extensive Darby activities in Coalbrookdale, widely acknowledged as the foremost iron-making concern in the kingdom. With his ear to the ground to pick up local talk of their activities and processes he would know that the Coalbrookdale lease was up for renewal in 1759, and that the landlords were in dispute with the Darbys over renewal conditions.

With astonishing confidence for a young and relatively unknown new arrival in the area, John then decided, no doubt with Brooke Forester standing behind him and with the New Willey Company well forward in its planning, to compete with the Darbys for the lease. Articles of agreement with the Darbys' landlords, John and Rose Giffard, were completed, though not disclosed, on 12 September 1757 for a lease to the New Willey Company to take effect at Michaelmas 1759, when the old Darby lease terminated.

There followed almost five years of litigation between John and Rose Giffard, and Thomas Goldney and Abraham Darby II for the Coalbrookdale Company, in a series of claims and counter-claims from which John and the New Willey Company were largely absent. The action ran its course through the Court of Chancery and into the Court of Common Law at Shrewsbury, where it was eventually resolved in a compromise agreement in the late summer of 1762.

The Coalbrookdale Company had no doubt been outraged when they heard of the Giffards' undisclosed agreement with John Wilkinson, whose only appearance in the dispute was at a number of informal meetings with Darby and Goldney where each side tried, and failed, to buy off the other. A later suggestion by John that the two companies should jointly run the Coalbrookdale works was rejected outright.[8]

It is interesting that at this early stage in his iron-making career, and at a time of much uncertainty in his personal affairs, John chose to take on the might of the Darbys rather than to cultivate them in a co-operative way. He must have known that his actions would alienate them and he must have believed, therefore, either that he had a good chance of acquiring the lease (of what was, after all, a prime iron-making site with its infrastructure already established), or that he would at least come out of the confrontation with something of value.

It is fascinating to speculate what that something might have been. Did he see the Coalbrookdale works as a possible alternative, or as an addition, to his plans for the New Willey Company? Or did he simply seek to divert the Coalbrookdale management from proper supervision of their works and hence slow down production whilst he was seeking to establish his company? Perhaps he knew that Abraham Darby II was under serious financial pressure at this time and had to rely on Thomas Goldney to bankroll the litigation. And perhaps he was able to apply pressure to the Coalbrookdale Company ,as a result of his actions here, to release to him on more favourable terms the old furnace at Willey on Forester land, which had been leased to the Coalbrookdale Company for years. It is of course an early example of that characteristic attitude that recurs so constantly in his later business life: the shrewd exploitation of a timely opportunity. In this case, to establish his name as a force to be reckoned with alongside some of the foremost personalities in the iron-making world.

Certainly, there was a huge task in front of his New Willey Company. The old furnace had been operating at marginal profit through shortage of water for the bellows for the past twenty years, and, as a consequence, Richard Ford and Thomas Goldney were prepared to release it. The furnace made only pig iron for the Bristol market, which is where John now found prospective partners with the necessary capital.

Edward Blakeway, a wealthy Shrewsbury businessman, joined John in this new venture. Blakeway probably brought in John Skrymster, also of Shrewsbury, and there was of course that shrewd Wilkinson touch in securing Brooke Forester as a partner, along with the land-owning Forester family's mining agent, William Ferriday. Eventually, six Bristol merchants took an interest in the business – providing the broader capital base the new venture required, and establishing a link between the Wilkinsons and Bristol at this date. This is the first appearance, too, of William Ferriday in the Wilkinson story.

The lease between George Forester and the New Willey Company was dated 13 June 1757 and was for forty-two years.[9] It included clauses giving authority to build new furnaces, to take large quantities of clod coal and ironstone from the Forester lands, as well as clay, sand and stone for building purposes, and to lay iron rails to transport goods both within the new mining and iron-making complex and across other Forester lands to the River Severn. It is a long and detailed document and provides long-term security for the company. It also describes John as 'Ironmaster, of Bersham', which establishes his known involvement in iron-making by that time, and indicates that his link with the Bersham furnace was retained. He was, in fact, the only ironmaster among the ten founding partners, which placed him exactly where he wanted to be: in direct control of the iron-making process.

A separate partnership deed for the New Willey Company partners was signed by all ten of them on the same day.[10] The starter capital of £16,000 was contributed in ten shares: Brooke Forester holding four; the six Bristol merchants two between them; and John, Edward Blakeway, John Skrymster and William Ferriday each holding one.

A precisely worded clause stated by what dates within the first year the proportions of this capital must be contributed. Others dealt with the keeping of, and access to, the company's account books and journals, the calculation and payment of an annual dividend, any unforeseen costs to the company, and compulsory attendance at the partners' annual meeting. Any disagreements were to be decided by a vote on the basis of the shareholdings, which immediately required a difficult degree of unanimity among the six Bristol merchants with only two votes between them. If voting fell evenly, Brooke Forester was to have a casting vote. Partners could sell their shares to existing partners after six months' notice in writing was given, but not to anyone else without full company consent. No new shares would be declared over and above the ten original shares. None of these restrictions, however, applied to Brooke Forester.

Two long clauses detailed how partners' shares might be acquired by the company if they failed to pay their proportions of capital by the due dates, or if they had used their shares as security for a debt, which was then called in. A major part of the document was thus devoted to circumscribing the partners' options and establishing procedures for any failure on their part to meet payment deadlines, or to agree. Brooke Forester's dominance was established and his special position recorded.

It seems that John, from the beginning, had seen that a break-up of the partnership was likely, perhaps even desirable, once the company had become established, and it is not difficult to see in this partnership deed how he might have engineered things to gain sole control of the company, which is in fact what happened. Much more difficult to understand is why, in a new partnership with Edward Blakeway (also of that company), only five months after the New Willey Company was established, he took another lease with authority to build a new furnace for Moreton Forge at Moreton Corbet, some 12 miles to the north of the New Willey Company works.[11]

It was an old manorial forge site long out of use, as the lawyer's recital of the lease shows, with the old forge pool, then a meadow, and the pool dam in need of repair. It had a good water supply from the River Roden and its own coppice woods, and came with three small messuages and a close of meadow in Moreton Corbet. John and Blakeway took the lease from Andrew Corbet on 17 November 1757, to run for forty-two years from 25 March 1758 for a rent of £26 a year, for the purpose of re-melting iron pigs and the forging of malleable iron.

It seems likely that, since the place was just a few miles north-east of Shrewsbury, news of its availability will have come first to Edward Blakeway, who would have known of the ancient Corbet family of Moreton Corbet. The site would then have been approved by John for its iron-making potential, even though some initial reconstruction and repair was required.

It is an important place in John's story because, alongside his newly established New Willey Company, this is where he first made iron in his own right. Up to this point he had been an ironmaster by virtue of his association with his father at the Bersham Ironworks, and it is interesting that in this deed he is described as 'Ironmaster of Wrexham', not Bersham, perhaps indicating that he wished to establish this departure. Where he resided in these years is not known. It is unlikely to have been his father's house at Plas Grono, which was 30 miles away.

It may be that the purpose of the Moreton forge lease was simply an insurance policy, to cover the possible failure of the ambitious New Willey Company project when it was not yet clear how the company, with its large list of partners, would work out. Or did he see Moreton forge running concurrent with the much larger New Willey Company and benefiting from that association, but remaining very much his own concern, using perhaps the New Willey Company production of iron pigs to make its own malleable iron? Certainly the lease of Moreton forge establishes a close working relationship between John and Blakeway, which was to mature and become important to both of them in the years ahead.

It seems clear that there was uncertainty on the part of both landlord and lessees in those early months of the New Willey Company, as their ideas and working practices began to form in late 1757 and through 1758. Nevertheless, in 1759 a second lease was signed between them.[12] It rehearses the details of the 1757 lease and refers to the high initial outlay from the partners in mining trials and new buildings and renovations before a steady income was on stream. The landlord, too, wished to be released from his agreement to supply all the building timber the company required, which was obviously decimating his woods and perhaps interfering with his sporting activities. Under the new agreement, the area over which the company could obtain clod coals and ironstone was extended to the whole estate, with the exception of one area already leased to someone else, and also excluding the Deer Park and any site within 500 yards of Willey Hall. The agreement was to run concurrent with the 1757 lease, cost the company another £800 a year, and consolidated their position.

Just before the second New Willey lease was signed, and with the detail of it no doubt by then established, John and Blakeway abandoned their interest in the Moreton forge. On 3 August 1759 they assigned the forty years remaining of that lease to a co-partnership of five local men who took over the improved site with permission to build a new forge with weirs, floodgates and associated water controls before 25 March 1761 – evidence of its continuing worth as an iron-making site and a vindication of Wilkinson's judgement in the first place.[13] But now he had a much bigger project in hand. In the next few years the New Willey Company became one of the most important iron-making concerns in the region, the site was improved and transformed, and little by little he gained complete control of it.

His sister Mary's courtship and marriage ran concurrently with these events, which also saw William return from school as a 17-year-old stripling to assist his father at the Bersham furnace in 1761. John was making an increasing success of the New Willey Company and, in contrast, young William had everything to learn. He had the advantage of having grown up in an iron-making family and he had Isaac to teach him, and Isaac would see there was much to be done quickly if William was to become a significant part of any family iron-making business. Isaac was 57 years old and the ensuing events indicate that he remained yet hopeful of a family business involving his sons, even though his own departure might be necessary to accomplish it. It is a remarkable commitment, but it may not have been as self-sacrificing as at first appears. There is evidence that he had seen new opportunities in South Wales and had already taken steps to secure a position there for himself.

In 1759, two years after his iron bellows patent and in partnership with eight other men, Isaac installed his blowing engine at the Merthyr Furnace, Dowlais. The wording of the agreement is interesting:

... and whereas Isaac Wilkinson of Plas Gronow Denbighshire both obtained a patent for his new invented machine for blowing furnaces, forges and other iron works by valves, Cyphons and both agreed with the other partners that they shall have the benefit of the Patent, in their said furnace. If any one furnace ... should make on an average more than 20 tons of metal per week during the first Blast then Isaac Wilkinson should receive the sum of £50 for every ton of pig iron so succeeding the quantity of 20 tons.[14]

Here are all the old hallmarks of Isaac: very aware of a forthcoming business opportunity and shrewdly carrying it to a favourable conclusion, in this case as a sleeping partner, since he was fully committed at the Bersham furnace at the time. His partners in the Dowlais enterprise were four Bristol and three Glamorgan merchants, and the same Edward Blakeway of Shrewsbury who was already in business with his son John. That, too, is significant because it suggests father and son remained close, although each had by now his own iron-making establishment. It could mean more than that. Isaac might have encouraged John to branch out on his own as a means of broadening the ultimate family base he wished to promote, even though John may not have been party to all the implications.

Certainly Isaac became extremely supportive of what he was to leave behind at Bersham. In 1762, acting for the Bersham Company, he entered into a price control agreement with John on behalf of the New Willey Company, and Abraham Darby II, acting for the Coalbrookdale partners, which fixed the price for the supply of engine parts to most markets, London to be excluded. The Darbys had obviously recognised the Wilkinsons as significant competitors in the iron trade when they signed up to this, and Isaac was moving ahead of them in another way at Bersham. He was again manufacturing cannon, and obviously encouraging John at New Willey to do the same. The Darby iron founders, because of their Quaker beliefs, never became openly involved in the manufacture of weapons of war. The Wilkinsons had no such reservations, taking the profits where they could find them, and so had a distinct advantage.

Another move seems to indicate that Isaac was thinking about the long-term future of the Bersham works immediately before he left. Tradition has it that in 1762 he handed over Bersham to his two sons, John, aged 34, and William, 18. There is no surviving document to record the terms of this transaction. It must be safe to assume, however, that John became the senior partner, perhaps with a commitment to bring William more and more into equal responsibility. For William continued to reside at Plas Grono after Isaac left, with all the appearance of being resident manager at what the brothers now called the New Bersham Company. What particular terms related to Isaac in any agreement remains unknown. Did he retain a financial interest? He moved from Bersham to South Wales shortly after this, but not before he had negotiated a new and improved forward lease with Richard Myddleton of Chirk Castle for the Bersham furnace.[15]

Under the new lease the area of the site was to include a further field, and authority was given for a cut to be made from the river to carry an improved water supply to the furnace – provided it did not interfere with the water supply to Bersham Mill. The length of the lease was extended to thirty years beyond the life of the then owner and a new access road was authorised, all this for 'the clear rent of two pounds of good and lawful money of Great Britain' to be paid every year to Richard Myddleton. The lease is dated 1 January 1763 but is to run from the previous 29 September, and is signed by Richard Myddleton and Isaac, described as 'of Plas Grono in the said County of Denbigh Iron Master'. It is persuasive evidence that he sought to support the future iron-making activities of his sons at Bersham after he had gone, and must call into question the view that he left in anger because of a quarrel.

In 1763, in partnership with John Guest of Broseley, Isaac leased land from the Earl of Plymouth in Merthyr Tydfil and erected a furnace there at what became known as the Plymouth Ironworks.[16] Four works were to become the mainstay of the important iron industry in the Merthyr area, and soon after he left Bersham Isaac Wilkinson had a founding interest in two of them. Did he continue to work towards a Wilkinson family of ironmasters? And did his hope of achieving it now include a centre of operations in South Wales?

By 1763 John and Edward Blakeway had been business partners for six years, and their association had survived, and apparently recovered from, Blakeway's bankruptcy in 1760.[17] How close they were as friends is not known, though it is likely they would dine together from time to time and perhaps entertain merchants and other wealthy individuals with capital to invest. Edward Blakeway was married to a woman well able to act as hostess on these social occasions. John was a widower and at this time unattached, indeed it was in this context that he met the wealthy sister of Blakeway's wife, a Mary Lee of Wroxeter, who in 1763 was 40 years old and unmarried. There can be little doubt that John's status as an unattached widower would create awkwardness in the circle that he and Blakeway wished to cultivate, but there may have been a more immediate reason.

The two sisters, on the death of their father, Thomas Lee, had received a moiety on his estate which was valuable in terms of land and property rather than income, and Edward Blakeway's wife had sold her interest to her sister Mary, presumably for ready money to pay creditors following her husband's bankruptcy.[18] It proved to be a neat accommodation in the event of Mary's marriage to John Wilkinson, her brother-in-law's business partner. Family property was sold to solve a family bankruptcy and as a result of the later marriage the property stayed in the family. It is rather too neat to be fortuitous, but if it was planned, were they all party to it? If not all, how many? And which of them?

41

No information survives about the courtship of John and Mary Lee. Was their marriage in 1763 entirely a marriage of convenience? Mary Lee was no mean catch. Why had she stayed unmarried for so long? Was she beautiful? Was her marriage simply an escape from spinsterhood into an exciting life she had glimpses of already, or did this young widower capture her heart? She brought substantial wealth to her marriage and that itself would be important for John at the time. He could not reasonably expect her to bear him children in view of her age. Did he settle for a consort rather than a woman, a partner to support him in his rise to power rather than a soul who could bring radiance and passion back into his life?

It is probable he found both. From what we know of their life together for the next forty years; from his pride at being able to leave her in charge at headquarters in his absence; and from the gentleness of his enquiries after her through his friends when they were apart, the marriage, no matter how it began, grew into a strong and caring association ended only by death.

He now had the reason, and the means, to find a new home and he took over a substantial three-storey, double-fronted Georgian house with a large, secluded garden on the sunny side in Church Street, Broseley, just across the road from Broseley Hall, where Edward Blakeway and Mary's sister lived. The house was called The New House at that time, later The Lawns, and had been built in 1727 by a local mine owner called Stevens. Although John may have rented it initially, he later took it on a ninety-nine year lease from the same George Forester from who he had leased the furnaces and mining rights for the New Willey Company.[19] In 1800 – by which time the focus of his life had shifted to his Castle Head estate in the north – he assigned the lease to the china manufacturer, John Rose, for 30 guineas a year, 'the said John Rose paying the whole of the window tax'.[20] But for more than thirty years it was his principal home, his headquarters and the centre of his business world.

His 7-year-old daughter, Mary, came back to live with him at this point and formed a strong bond with her new stepmother and half-cousins, particularly Elizabeth Clayton, and thus ended her father's years of grieving and loneliness. This happy transformation of his personal life came at an opportune time. His acclaim as a young ironmaster was growing; he was obtaining ever more control of the expanding New Willey Company; and he was now the senior partner in control of the New Bersham Company. There was much to do.

4

WEALTH AND ACCLAIM

In 1763 France was defeated by Britain in the Seven Years War. An important factor in this victory was considered to be the superior quality of the British naval cannon, though even at that date it was said that our own naval gunners were more afraid of their cannon exploding on ignition than they were of the fire from enemy guns. The superior reputation of the British cannon, however, persuaded the French government to send to England in 1764 a young engineer from Lyons, called Gabriel Jars, to report on iron-making methods around the country and to try to establish if it was the quality of the English iron alone that enabled them to make better guns.

The Wilkinsons, by now important ironmasters in the Midland iron-making area, would have known of this visit and its purpose (though a meeting between them and this Frenchman is not documented) and they would have identified the opportunities the French interest offered, following the end of hostilities, for an increased sale of naval cannon. Isaac, of course, had already identified the war as such an opportunity and the Bersham works, unlike those of its Quaker competitors, had been involved in the manufacture of guns for some years. Gun-making was now included at the New Willey works alongside the production of engine cylinders for the old Newcomen engines, and it is clear that the main thrust of the Wilkinsons' iron manufactures in the following years was towards the supply of cannon for government ordnance. This resulted in steady improvements in the processes as they gained more experience, with good profits accumulating from the sales.

John's first patent application in 1765, however, shows his interest in a very different direction. A summary of the subject matter comes as something of a shock: '... medicated baths – constructed on frames for floating on water; floats made of cork in the form of a seaman's waistcoat, to be used to prevent drowning.'[1]

What was he doing here? The application points to his broader involvement with Samuel More and the Royal Society and the contact this would bring him with the social activities of the day, of which 'taking the waters' would be one. Did he, with his shrewd business brain, see an opportunity to market a device that would gain him access to a wealthy elite grateful for this support of their physical infirmity, and whose capital might therefore be attracted the more readily into his enterprises? That of course is speculative, but it was certainly a time of growth, expansion and investment in his affairs; there is evidence that he attracted capital from the Bristol area, which included Bath, and he took every opportunity to advertise his products.

His first contact with Matthew Boulton the following year further illustrates this. A certain John Florry of Birmingham was obviously an influential and very satisfied Wilkinson customer, and had suggested he approach Boulton with a view to extending his sales of iron castings. John's reply to him, dated 5 December 1766, included the following:

> ... As I have not the pleasure of being acquainted with Mr Boulton my writing to him on the business you have been so kind as to recommend me in, would not in my opinion have that weight as your application on my behalf, for any castings he may have occasion for. I should have a particular

Samuel More, secretary of the Society of Arts, friend, confidante and supporter of John Wilkinson.

pleasure in doing business for a Gentleman of such distinguished merit – as by that means I might form an acquaintance with a Genius, that might in future afford me great satisfaction on many accounts. I need not I hope take notice to you – that you run no risque of disgrace in your recommendation of my abilities to serve him well. I will undertake to promise that what ever he may want will not, or cannot be better executed by any one. As to the prices – I can say nothing to 'em until I was to know the nature of the articles wanted – but this may be observed in general, that in all engine work the Dale Co. and I observe one rate and rule ...[2]

It is a letter which rings with confidence and self-possession, recognising the importance of this new customer and using a mixture of flattery and reassurance whilst retaining that shrewd business acumen that marks the Wilkinsons. It confirms, too, that following Isaac's departure his two sons had cannily retained the price control agreement with the Darbys, set up in 1762. The original letter contains an additional hand-written forwarding note signed by John Florry and addressed to 'Mr Matt. Boulton', which starts off with the following:

Sir,
    I beg to refer you to the above, also to asure you that I believe there isn't one of the trade understands their business better than Mr Wilkinson ...[3]

Unfortunately, the immediate follow-up contact with Matthew Boulton is not recorded, but there can be little doubt that John Florry here promoted the single most important piece of business in John's rise to power, and the absence of any further documentary evidence for a growing link with Boulton in the next few years is very tantalising.

Samuel More also begins to draw attention to improved methods at the Wilkinsons' works during this period and from his base in London, and, in his journeys far and wide across the country looking for new inventions and ideas, he was extremely well placed to promote his friend. As seen here, John had obviously devised a modification and improvement to Isaac's iron bellows and More was attempting to explain this to his mentor at the Society of Arts, Dr Lewis:

... There can be no objection to my sending you the best account I am able of Mr Wilkinson's improved Water Bellows, as he wanted much to see you to explain it himself to you and if you find any objections to it I shall be glad to hear them that I may let him know your sentiments which I am sure he will pay great regard to. What first set him about this improvement was that by no bellows yet worked by water could there ever be obtained a cubic foot of air by a cubic foot of water expended in working them. This he imagines he has remedied, by contriving to convey the water out of the vessel ... by means of a syphon ...[4]

The close friendship with More continued through the next decade and beyond, until More's death, in fact, in 1797. Indeed his promotion of John went beyond the support of his works and ideas to an enthusiasm for the man himself, as later events will show.

These years were a period of personal and business consolidation for John. The Lawns at Broseley became his family base and what he increasingly came to refer to as his 'Headquarters'. Here, with Mary Lee now as hostess, he entertained his business friends in a home which was comfortable and commodious without being grand, and which was geographically central to his affairs. It was only a mile or two from his works at New Willey, about 30 miles to Bersham – which enabled him to travel there and back in a day and thus offer support to young William's management of affairs there – and well within a day's return journey the other way to Birmingham and the world of Matthew Boulton and his friends, and also to Bradley in Bilston, between Birmingham and Wolverhampton, where he had great plans afoot. In 1767 he bought the estate and the Manor of Bradley and began to build a new blast furnace there. It was to become another of his great iron-making centres and a place with which he was to be associated until his death.

His daughter, Mary, was now growing up at The Lawns, moving into her teenage years and drawing closer to her stepmother, perhaps as the daughter that Mary Lee herself was never able to bear. A bond was also established at this time between John and his wife and daughter, which was to have a powerful effect upon their later lives. Subsequent letters to James Watt show how much John enjoyed his Broseley home of which his womenfolk were such an integral part; their greetings and salutations always included with his own at the end of his letters.

Sometime before this he had become acquainted with the architect Thomas Farnolls Pritchard. The memorial to his first wife in Wrexham church is by Pritchard; it may not have been erected until some years after her death in 1756, though before his second marriage in 1763. There is no documentary support for the start of this association but it continued through the early Broseley years as witness a piece of architectural evidence at The Lawns, suggesting Pritchard was an intimate in the Wilkinson household at that time. A monumental marble chimney piece erected by him still dominates the dining room there. Ten years later, of course, Pritchard became the architect of the great Iron Bridge over the Severn, a scheme that John sponsored and promoted from the very beginning.[5]

Isaac Wilkinson was now living in Bristol. A reference from the Merthyr Tydfil area dated 2 December 1768 lists him as 'Gentleman, of the City of Bristol'.[6] Whether he ever lived in the Merthyr area of South Wales after his move from Bersham, or whether he moved straight to Bristol and conducted his South Wales investments from there in semi-retirement as some sources suggest, is unclear. He turned 63 or 64 years old in 1768.

Glimpses of Isaac and his relationship with his children can be seen in the references that come from his son-in-law Joseph Priestley, who stayed with him at his Bristol home on occasions. That there had been a serious breakdown about this time in relations between Isaac and his eldest son, John, is confirmed in a much later letter (1796) from Priestley in America informing John of the death of Mary, his wife and John's sister. He says of her: '... She always warmly took your part and would never believe your father's account of your using him ill.'[7]

It is an important glimpse of that father–son relationship, and it is significant in the context of Mary's marriage to Joseph Priestley, at which ceremony her father was notably absent. Again, it is all tantalisingly incomplete. It gives no date, no details and no indication as to whether the relationship improved, or further deteriorated. Perhaps John, in his rise to power, had turned aside from any hopes his father still held of heading a Wilkinson dynasty of iron founders; had dismissed the South Wales venture as too distant from, and irrelevant to, his own plans in the Midlands. And if Isaac's South Wales investments were an attempt by him to provide for his own old age in the face of his dismissal from the plans of his eldest son, they, too, were unsuccessful.

Writing in 1795 in his *A Description of the Country from Thirty to Forty miles round Manchester*, Aikin says that Isaac Wilkinson's Old Bersham Company 'proved unsuccessful, partly in consequence of an expensive scheme to convey a blast by bellows from a considerable distance, to the works by means of tubes underground',[8] stating that it was instead his son John who adapted the device and made it succeed after Isaac had left. In the next ten years the New Bersham Company under John and William began to market blowing engines for blast furnaces with increasing success, and that alone would have been cause enough for a rift with Isaac, particularly if he sought, and was refused, a share in the profits to lessen his financial difficulties in South Wales.

Isaac was ultimately involved in three major iron-making ventures in the Merthyr Tydfil area, none of which were a financial success for him, yet which made fortunes for later ironmasters after he had withdrawn to Bristol.[9] So his instincts were again correct, even if his operations on the ground were lacking.

Yet another reference cites the underground air pipes as a contributory failure, this time in connection with the Dowlais and Plymouth furnaces:

> ... at a considerable distance from this furnace there was a water-wheel which acted as the motive power to a large bellows, supplying the furnace with blast. The blast again was conveyed through a long clay pipe of a very frail character. The whole thing soon collapsed and Wilkinson retired from both Dowlais and Plymouth ...[10]

It is likely that Isaac's ultimate financial embarrassment, however, was the result of a ruinous and protracted court case in which three master colliers from the place of his third and final speculation, the Cyfarthfa ironworks in Glamorgan, claimed for 'a considerable sum of money' owing to them for coal produced under a contract between them.[11]

Isaac's financial salvation in his declining years in Bristol is likely to have come from John Guest of Broseley, an early partner in his South Wales speculations who made a financial success of the Dowlais Ironworks in the years after Isaac left. Almost twenty years later, Guest is described as 'a creditor of the deceased' in letters of administration over Isaac's possessions granted to Guest's son, Thomas Guest, following Isaac's death. In those same letters, Isaac's surviving children, John, William, Mary and Margaret, renounced all interest in his estate. He died intestate.

That Isaac retained some dignity and status in his declining years in Bristol is clear both from the directory listings of him from time to time,[12] and from the notice of his death in 1784.[13] His two youngest daughters left his Bristol home to be married, both to men of some position and standing: Sarah in 1768 to Thomas Jones, a surgeon and apothecary in Leeds, and Margaret in 1771 to Thomas Parkinson, a glass seller in London.

There is one further piece of evidence relating to John's relations with his father in his declining years. A letter from Matthew Boulton to James Watt, dated 23 September 1781, contains the following:

> ... I was prejudiced against W[alker] before I saw him, but that is now vanished and don't think we shall have any cause of complaint. It was Le Founder [i.e. John Wilkinson] that grounded my prejudice, but as I learned from old W[alke]r that he had been appointed an arbitrator between Le Founder and his Father and that after the award was made Jno. would not abide by it, but brought the affair into the King's Bench again, when Lord Mansfield confirmed the award ...[14]

The patchy evidence certainly points to both the financial failure of Isaac Wilkinson in his South Wales ventures after he left Bersham, and to a quarrel about this time with his eldest son John. It seems likely the one is related to the other.

There is, however, good evidence for another important yet diverging focus at this time in John's life: his continuing contact with the Castle Head area of Westmoreland, just 1 mile downriver from his old home at Wilson House.[15] The information comes in a treasure box of redundant documents purchased some years ago in a solicitors' clear-out in Liverpool.[16] The documents included a long series of lawyers' abstracts relating to the later litigation over the will and the disposal of the Castle Head estate following John's death, and rehearsing deeds involving the purchase of land within the estate dating back to 1704.

The secondary sources on John, mostly published many years after his death, repeat the information that he purchased the rocky hill of Castle Head in 1765. That information is not supported by this bundle of documents, which indeed make clear that he did not complete that purchase until 1778. They do, however, show a link between John and the purchase of a property lying adjacent to the hill in 1761. The property is described as follows:

> all that messuage or dwelling house at or near Lindale commonly known by the name of The Pot House ... and two parcels or arable meadow or pasture adjoining lying together in one enclosure known as ... The Intake and Sandy Bridge Parrock ... [and the deed is between] ... William Canney of Lindale ... yeoman of the one part and Thomas Waller late of the same place but now of New Willey Furnace in Shropshire, house carpenter, of the other part ...[17]

The document is important because it proves the link between the Lindale family of Waller and the Wilkinsons, in the context of an early land purchase at Castle Head. It seems that perhaps the Wallers went with Isaac from this neighbourhood to Bersham in 1753, and subsequently came to work for John at his New Willey Furnace. Had they prospered? Were they homesick? Was this a move to return to the place of their roots? Or, conversely, did John make this purchase through a loyal and trusted servant who may well have been homesick and whom he intended to install in the property until other plans were ripe? Finally, did he provide Thomas with the purchase price of £107 3s to do it? This would have been a huge sum for a house carpenter to find at that time.

Another abstract in the box of documents referring to part of this land raises more questions. In Agnes Waller's will, dated 23 December 1786, the field called The Intake, which had passed to her following Thomas's death, is specifically left to Mary Wilkinson, wife of John Wilkinson. There is no reference to The

Pot House, or to Sandy Bridge Parrock, just the one field, but it would be unusual for a lowly family like the Wallers to leave land to their masters, the Wilkinsons, unless there was a bond or agreement between them from an earlier date. The fact that the Wallers left their home to follow the Wilkinsons from Lindale to Shropshire, came back again and continued to serve the family until they died does indicate an enduring and close relationship within which such an arrangement might be made.

By the early 1770s the name of Wilkinson was attracting ever more attention as a supplier of good quality iron products in a market, until then, dominated by the Darbys of Coalbrookdale and their Quaker partners in this Midland sforefront of the iron-making world. Little is heard of William Wilkinson at this time. John is the dominant force, though William's management of the profitable gun-making works at Bersham was making steady contributions to their increasing wealth. In 1770 William was 26 years old. John was 42, and there are important differences between the two brothers as a consequence of this sixteen-year gap.

They can have shared little as children. Apart from any comradeship John found with his younger brother Henry, he must have had a lonely childhood at Backbarrow, where he followed ideas and decisions made largely without discussion or consultation. William, on the other hand, went through childhood as the only boy in a brood of sisters much closer in age – one older, two younger – and although he would perhaps be dominant in their play, that would also involve responsibilities of negotiation and justification. Moreover he would be closer to Isaac. John was absent from the family fold for almost all the five Wilson House years, and for those early years at Bersham, by which time he was married. William was with his parents throughout and Plas Grono was the place where William grew to manhood. It was his home in a way it was never John's, and he had an intimacy with the Bersham furnace and the people who worked there which went back into his childhood. As manager, however, because of his youth he would need to prove to the people who looked increasingly to him in John's absence that he was able to make good decisions about staff and markets and production processes to make the business prosper. For him, too, this was a period of consolidation.

John, meantime, was experimenting with iron-making ideas and machinery in a practical way, at New Willey and at his new Bradley furnace, within the various systems producing iron products there, in a questioning, searching manner that was typical of Isaac. It was a restless, questing work ethic, totally committed to iron as a medium, constantly seeking better processes, better products, better marketing – all of which lead to more power, increased wealth and greater acclaim.

Stockdale[18] quotes from a letter John wrote to Matthew Boulton, which not only illustrates this point but also makes clear that in the six years since their initial contact through John Florry, the two men are now intimate enough to be sharing information and ideas:

> Bradley, October 11th, 1772.
>   I am happy to acquaint you that I have at last succeeded in using coal in my furnace. The coal is got on my estate, and answers well. The produce of the furnace weekly is now twenty tons instead of ten as formerly ...

Charcoal, in the quantities required for iron furnaces, as industrialisation and therefore demand dramatically expanded, had become increasingly expensive and difficult to obtain. The Darbys had used a crude form of coke as a furnace fuel since earliest days but it was never entirely satisfactory. John's success in 1772 seems to have come after extended trials with other fuels, one of which had taken him back to his roots at the Backbarrow Furnace, where Fell says he again experimented with peat: 'In 1770, John Wilkinson carried out some experiments with peat charcoal at Backbarrow, and the furnace was placed at his disposal, but whether successful or otherwise is unknown.'[19]

Again, Fell does not disclose his sources, but his information is interesting because it indicates that there was still a Wilkinson link with the Backbarrow furnace more than twenty years after Isaac left, sufficient for John to be allowed the use of it. John's name alone by this time might have been enough to grant him this access in his own right of course. The other important point is that here he is again in the north, not far from Castle Head, for what must have been an extended stay.

The greatest breakthrough in this period was in the Wilkinson method of producing cannon. Earlier practices, and probably those used at first by Isaac, cast the heavy cannon barrel as a crude iron pipe and then bored out the inner surface with a rotating cutting tool to clean and finish it. The problem was that the cutting tool jumped about as it cleaned out the core due to inconsistencies in the hardness of the iron and faults in the cast metal, which meant it was near impossible to achieve anything like a true cylinder for the inside of the barrel. It is not difficult to see why cannon barrels had a tendency to explode in use.

John developed a totally different approach to this process. First he cast the barrel solid to eliminate faults, which invariably occurred in the thinner casting when using a core. He then introduced the solid barrel and the cutting tool to each other by an improved and much more rigid system of slides, and instead of rotating a cutting tool into the metal, he rotated the barrel and used a rigid cutting bar against which the turning barrel moved:

It is difficult to overestimate this invention as it was really the introduction of the principle of slideways to large machine tools. Up to this time the object to be bored, already cast to nearly the intended size, was mounted on a trolley which was drawn on rails towards the rotating boring head. The cutters took the path of least resistance and although any section might be circular, the hole was not necessarily cylindrical. Wilkinson dispensed with the core and cast the workpiece solid to eliminate blowholes. It was then rotated in bearings and the cutter was fed into it along a straight, rigid bar by means of a screw. Almost before the patent was taken out, Wilkinson modified the mill for engine cylinders ...[20]

The patent application was number 1063, dated 27 January 1774, 'for a new method of boring guns and cannon'. It was not long before the French heard about it, probably through Matthew Boulton's extensive network of trade contacts on the continent, and in July and August of the following year an infantry brigadier in Louis XVI's army, Marchant de la Houliere from Perpignan, came to England. His alleged purpose: 'to investigate whether the superior quality of English coal and the peculiar nature of English iron ores constituted the reason why they were used together with such a great measure of success in that country to make cast and wrought iron.'[21]

He was introduced to John by Matthew Boulton and Samuel Fothergill. For the next few days John was his host and offered him every opportunity to see his own works and those of his friends. It did not take de la Houliere long to focus his attention on the manufacture of cannon. He was an informed and observant visitor, with recent experience of attempts to smelt iron with coke in France:

On August 19th at Bersham foundry in Denbighshire in the principality of Wales ... 80 cwt. of cast iron were placed in four reverberatory furnaces in my presence. In one of them was placed half a broken cannon which might have weighed 15 cwt., and 5 or 6 cwt. of crude pig iron which had been smelted some time previously in a blast furnace using coke. In another ... was placed a damaged forge hammer ... To this was added some of the same iron referred to above in order to bring up the charge of each to between 20 and 21 cwts. It was 9.30 a.m. by my watch when all four were charged. By noon these masses of iron were in a state of complete fusion. The plugs were withdrawn from the tapping holes, and I saw a 32-pounder cannon run into the mould ... Cylinders from fire or steam engines are melted up and cast again in the same manner ... These reverberatory furnaces are heated with raw coal straight from the mine and are stoked for two hours before they are charged with crude or scrap iron. This is not done until the vault is white hot, and then pieces of cannon, anvils, hammers, old scrap iron castings or bars of pig iron are put in, and within two hours and a half the whole consignment may be seen passing through the pouring gate to be cast into the desired shape.[22]

It is an immediate and very detailed description from a man who understood iron making, and provides today better information on the Wilkinson works at Bersham at that time than any other surviving document.

The second part of Marchant de la Houliere's report is given in the third person, placing immediately a distance and a veil between himself and his subject. Why did he do this? Was it deliberate? Is he hiding something? As the guest of John at Broseley he makes clear his purpose, which is to bring to France 'an experienced man' to establish the new technology there, and says this was a sudden inspiration which came to him in England. But he is a military man, and the improved cannon is what he really wants. His veiled approach to John meets with a shrewd response. John sees at once beyond Marchant de la Houliere's nods and winks to the real masters in France and instinctively senses the problems. He tells the Frenchman he is interested but he would need assurances; a guaranteed market for at least twelve years with freedom to export any surpluses anywhere, solid guarantees of payment and exemption from taxes on coal. De la Houliere had not been expecting anything like this and went to sound-out young brother William. He met a very different response:

> Next day I was at Bersham foundry which is managed by his younger brother … I gave him an account of my conversation with his brother. He appeared to me to agree with all the views expressed, and had a great desire to see France … I have reason to think that this Englishman would not be long in arriving …[23]

He was right. The following year William went to France. His brief was to build, with the help of a French engineer called Pierre Toufaire, a new state ironworks and cannon foundry at Indret, an island on the River Loire below Nantes. He built a copy of the Bersham works, which had been his responsibility for fourteen years, complete with gun-boring lathes and reverberatory furnaces. The first cannon were cast there in 1779. By then France and England were at war.

The decision that took William Wilkinson to France was made in 1775 however, and it is worth examining his circumstances at the time. He was 31 years old, a successful young ironmaster but very much in the shadow of his elder brother, at an age when he would feel the need to make his own mark on the world. The French terms would have to be good enough to match his income and status at Bersham, if he was to replace the one for the other, and they were.[24] But was that in fact the position? Did he decide to abandon the Wilkinson empire at a demanding moment in its development, and without regard to the consequences, for something that suited him better at the time? Attractive as the French prospect would be as a focus of style and fashion to a young bachelor, quite apart from the salary and status, it is clear that William left with John's blessing and there was no rift between the brothers at this time.

John's support of William's inclination to go to France begs some questions, however. John had recently supplied a cast-iron cylinder 'bored to truth' to replace the cylinder on the old experimental Kinneil engine which James Watt, with Matthew Boulton's help, had salvaged out of Roebuck's bankruptcy in Scotland, and which Boulton and Watt had re-erected at the Soho works in Birmingham as a prototype new engine. Because of Boulton's trading network of contacts, the talk of Watt's new engine came ahead of its production. But the great association between the three men was beginning and a new emphasis for cylinders as opposed to guns was emerging in John's priorities. There would have to be changes at Bersham.

There is no doubt, too, that John was looking to the continent as an extended market for his iron products. Locating William in a post of high status in France provided an advance listening post for potential markets, which could greatly benefit the Wilkinson business, and there is later evidence that it certainly did. This French connection became an important episode in the Wilkinson story and we shall return to it later, but in 1775 the iron-making world of the Midlands was waiting to see if Matthew Boulton's advance advertising of Watt's new fire engine was going to fulfil its promises. There was excitement and expectancy in the air, and the first working engine had not yet been built.

5

THE NEW STEAM ENGINE

By 1775 talk of James Watt's new, improved fire engine had been a subject for speculation for almost ten years. As instrument maker at Glasgow University in the early 1760s, Watt had been given a model of a Newcomen steam engine to repair, which at first he was unable to do to his satisfaction. The brainwave for a separate condenser came suddenly in 1765 and he built a working model of an improved engine which excited attention.

He did not himself have the resources to build a prototype, but Dr John Roebuck, a wealthy industrialist from Birmingham with interests in Glasgow, heard about Watt's engine and a friendship developed between the two men. Roebuck subsequently paid off debts Watt had undertaken, in an attempt to develop the engine himself in exchange for a two-thirds interest, and provided the resources and encouragement for Watt to build a working model on Roebuck property at Kinneil.

Because of the demands on Watt's time, by then as a busy canal surveyor and engineer in Scotland, progress was frustratingly slow and the new model was not completed until 1769. An experimental working engine followed with an 18in cast-iron cylinder supplied by Roebuck's Carron Ironworks. It could never be made to work satisfactorily, but Watt was nonetheless encouraged to apply for a patent, which he was granted for a period of eight years on 5 January 1769 and in which he set down his specification and the principles on which the new engine was based.[1]

Richard Crawshay, the outstandingly successful ironmaster of Cyfarthfa, was a close friend of John Wilkinson.

By 1770 Roebuck was in serious financial difficulties. He knew Matthew Boulton and had already tried to interest him in a partnership in one of his enterprises, but Boulton had been fully committed with his then partner John Fothergill to the building of his new manufactory at Handsworth, the highly acclaimed Soho works.

Following a visit to London on canal business in 1768, Watt had returned to Scotland by way of Birmingham and had stayed with Boulton at Handsworth. Here he saw the sophistication of the new Soho works. He had been the guest of that informal gathering of scientific men started by Boulton and his friends in 1766 and known as the Lunar Society, so called, it seems, for no other reason than their need of the full moon to light them home after meetings. During his stay in Birmingham, Watt had been hugely impressed both by the friendship and hospitality extended to him by these men, and by their enthusiasm for his invention. It is not known if he met John at this time.

The possibility of Boulton being admitted to the Roebuck-Watt partnership had been mooted earlier and it was hardly surprising that, in the event of Roebuck's subsequent failure, Watt and Boulton, who had by then established a liking and respect for each other, should look for a way forward. Boulton was tremendously enthusiastic about getting the engine into production, but cannily held off Roebuck's initial proposals knowing he was almost bankrupt, and waited eventually to deal with his creditors. Watt, never happy with these financial manipulations, was torn between loyalty to Roebuck and his recognition of Boulton's Birmingham as the environment in which he might best build his engine.

In the period before Roebuck's ultimate financial collapse in 1773, Watt wrote a series of letters to Boulton and his new friends in the Lunar Society, particularly to Dr William Small.[2] They are extremely illuminating as to his state of mind at this time and show him by turns anxious and despondent, with a poor view of his own abilities, his health affected by worry, aware of his own precarious financial position and needing to carry on with the canal survey and engineering work to keep his family in bread; and hugely frustrated that this near full-time commitment kept him from further experimental work on the engine. They also indicate a restless probing intellect, which, frustrated in the engine work, turned to what he called 'gimcracks'. One such gimcrack was a notable improvement to his survey equipment in the form of a new micrometer, which he discussed with Small. As Watt's difficulties dragged on through these years he nonetheless retained modesty and a generosity of spirit, perhaps born out of a faith that his new Birmingham friends would somehow find the way forward.

James Watt, whose formidable combination of genius and diligence was greatly respected by Wilkinson.

At the meeting of Roebuck's creditors early in 1773 Boulton asked Watt to act as his attorney, which must have been a trial all round for Watt's loyalties, but his handling of the meeting and its consequences was masterly. First he discharged Roebuck from all sums that were owing or outstanding under their partnership, 'in consideration of the mutual friendship existing between Dr Roebuck and myself and because I think the thousand pounds he has paid more than the value of the property of the two-thirds of the invention'.[3]

Next, because the Kinneil experimental engine was astonishingly valued at nothing by Roebuck's creditors, Watt took it as his property, dismantled it and shipped it in pieces via London to Soho. This was the first positive decision by Watt to show that he now knew his destiny lay in Birmingham. The final consequence of the resolution of Roebuck's failure was that Boulton acquired the two-thirds share in the engine in exchange for an amount of £1,200 owing to Boulton and Fothergill. Fothergill did not want to carry forward a financial commitment to the invention so Boulton next paid him out for his share of this sum.

A consequence of Watt's own generosity towards Roebuck was further serious strain on his own finances. With his wife heavily pregnant with her fifth child he felt obliged to undertake a new canal survey, which took him to the wilds of northern Scotland between Inverness and Fort William. It is a measure of the man that at this difficult point in his life, instinctively aware that massive changes were imminent and weighted down by uncertainty and stress, he completed the detailed and accurate survey work over rough and difficult ground that was so praised by Telford (who in 1801 built this section of the Caledonian Canal).

Yet tragedy was now about to enter Watt's life. In that wild country, several days' journey north of Glasgow, he received urgent word that his wife was ill and not expected to live, and he should come home with all speed. He was met near Dunbarton by a family friend from Glasgow, Gilbert Hamilton. 'By his black coat and his countenance I saw I had nothing to hope.'[4] He couldn't face his own home at first because he 'feared to come where I had lost my kind welcome ... in her I lost the comfort of my life, a dear friend and a faithful wife'.[5] A letter to Dr Small in Birmingham a little later emphasises the misery of this time:

> I know that grief has its period; but I have much to suffer first. I grieve ... I am left to mourn ... I had a miserable journey home, through the wildest country I ever saw, and the worst conducted roads; an incessant rain kept me for three days as wet as water could make me. I could hardly preserve my journal-book ...[6]

His wife's death left Watt responsible for their two surviving children, the elder only 6. His Glasgow friends, particularly the Hamiltons, looked after them when he resumed his survey work as he had to do to provide for them. It was in this melancholy period he met Mrs Hamilton's sister, Ann McGregor, who was to become his second wife and whose subsequent letters show the important role she played in helping him adapt to his new life in Birmingham. Small, too, offered tremendous sympathy and support at this time and the two men developed an intimate relationship of trust and confidence unmatched by any other in Watt's life:

> I have lost much of my attachment to the world, even to my own devices ... I long much to see you – to hear your nonsenses and to communicate my own; but so many things are in the way, and I am so poor ... I am heartsick of this country; I am indolent to excess ... I tremble when I hear the name of a man I have any transactions to settle with ...[7]

It was Small as much as Boulton who finally persuaded Watt to break his ties with Scotland and follow his engine to Birmingham. He arrived on 31 May 1774, but sadly this sustaining friendship was cut short by Small's sudden death of ague in the following year. He and Boulton had persuaded Watt that since six years of the patent had expired without a single working engine being built, he should take counsel's opinion as to how the period of the patent could best be extended. The advice was to obtain an Act of Parliament rather than a new or modified patent and Small had drafted the petition for the bill shortly before he died. It met with strong criticism in the House of Commons when it was introduced on 23 February 1775. 'Violent opposition from many of the most powerful people in the house' claimed that the period of twenty-five years applied for was too long and the wording to prevent possible infringements too sweeping. [8] The shrewd mind of Matthew Boulton with an eye to future profits is writ large in these clauses, and also in the wording of the memorandum circulated by Watt to members of the House in which he seeks to justify his position:

> The inventor of these new engines is sorry that gentlemen of knowledge and advowed admirers of his invention should oppose the Bill by putting it in the light of a monopoly. He never had any intention of circumscribing or claiming the inventions of others; and the Bill is now drawn up in such a manner as sufficiently guards those rights and must oblige him to prove his own right to every part of his invention which may at any time be disputed.

Matthew Boulton, an entrepreneur able to recognise ideas and opportunities, and keen that his success should be recognised.

The bill was read for the first time on 9 March 1775. Watt was in London for most of its passage, supported by Boulton during the critical stage in May, but the bill was steered safely through Parliament and received royal assent on 22 May 1775.[9] The question as to how far this Act prevented other inventive minds from working on further improvements to the new steam engine for the next twenty-five years continues to be asked to this day.

During the year leading up to the passing of the Public Act Patent, Watt, now living in Birmingham, had been making further experimental modifications to his engine set up at Soho. At some point there will have been discussions with Boulton about the quality of the engine's Carron cylinder, which Watt, by his constant and systematic engine testing, had identified as a serious limitation against further improvement. The need of a better cylinder was urgent. Boulton, because of his status and as a regular Wilkinson customer for castings already, would know about the new cannon-boring experiments ahead of their general publication and might even have persuaded John to extend the technology immediately to cylinders. It would be entirely typical of John, of course, to have identified the excitement focussed on Watt's engine as a huge potential market for high-quality castings. The engine, in fact, was the catalyst for John's application of the new, improved cannon-boring technology to cylinders.

Certainly, by early 1775 John was working on a new cylinder for Watt, which was delivered to Soho in April. It is from these beginnings that the important cooperative association of the three men grew: Matthew Boulton, wealthy entrepreneur and highly respected small goods manufacturer with a wide trading network; James Watt, inventor and civil engineer, his new engine yet to be proved; and John Wilkinson, ironmaster and a practical engineer himself with a growing iron empire and reputation. Did they have one great unifying purpose: to make money?

In the summer of 1775 Watt, now installed in Matthew Boulton's old Birmingham house at 1 Newhall Walk, went back to Glasgow to collect his children, and perhaps to see Ann McGregor. He and Boulton had obviously discussed the contractual terms of their business association and in a letter from Glasgow dated 5 July, distanced from the intensity and excitement at Soho and where he had leisure to think about this and discuss it with his friends, Watt rehearses the agreed terms to Boulton – 'as you may have possibly mislaid my missive to you concerning our contract'.[10] It is possible that Boulton, with the two-thirds share of the engine in his pocket, had been slow to finalise the contract with Watt and that this was a murmur of concern from Scotland. This is supported by another vibration the following year when Watt returned again to Glasgow, to marry Ann McGregor, with the contract still not formalised.

Her father insisted on seeing the contract document before he would agree the marriage settlement. It had not then been executed, but with a huge amount of goodwill on both sides Boulton reassured the old man and promised him sight of the document when his lawyer returned from London. It is an interesting exchange, the more so since the executed deed has never been found, though it is perhaps unwise to read too much into that. Boulton and Watt from earliest times had each shown a respect and admiration for the other's very different personality and skills, and their continuing relationship as business partners shows, too, that each retained a steady trust in the other's integrity.

Watt was in direct contact with John in 1775 following his delight at the improvement the Wilkinson cylinder had made to the experimental engine at Soho. Plans were soon afoot to build the first working engines for sale, one at the Bloomfield Colliery near Tipton to pump water, the other at John's New Willey works to perform a much more complex operation. It was to be harnessed to his iron bellows to blow the furnace. It was an anxious time. Potential purchasers were waiting, and watching. Each engine had to succeed. James Watt was nervous. John wrote to him reassuringly:

John Wilkinson to Mr Watt at Soho, near Birmingham. Broseley 17th Aug 1775.

Dear Sir,

I have just rec'd your favour with the several drawings. Some gentlemen in this neighbourhood having an engine to erect – I have advised them to build it on your plan. One of the partys that pretends to understand these things has promised to accompany me to Birmingham the first time I come over and as I wish to see you before you go northwards I shall fix him for Monday next and purpose to call at Soho in our road to Birmingham that evening. I shall then mention to you what occurs to me in the business you have so well described in your letter and then forward the instruction to Bersham. I wish to do all in the best manner and to start fair. Let us only succeed well in these first engines, particularly in mine, and I will venture to promise you more orders than will be executed in our time.

Am glad you think the cylinder for Mr Bailley & Co will do – am preparing a machine in the new way to finish them with greater truth. If y will not answer a perpendicular x shall be tried. In short nothing shall be wanting that is in my way which can promote and facilitate your engine. Our time in this world (at best) is but short and we must be busy if you intend that all the engines in this Kingdom shall be put right in our day. I have had some thoughts on this matter, and am of opinion that if you dare undertake the drawings I will provide the castings. Practice will make us perfect, and a score of engines one year hence will be dismissed with more ease than one at present.

I beg my compliments to Mr Boulton and am Dear Sir, Your most obedient Servant, John Wilkinson.[11]

It is an informative letter. There is a certain formality of address here showing the two men are not yet on familiar terms, and an unmistakable politeness from John towards a man whose intellect and invention command respect and whose drawings he cannot match. Already John is promoting the engine, and seeking to bring to Soho to see it 'one of the partys that pretends to understand these things' with a view to procuring an order. Watt described certain modifications in an earlier letter, which John wanted to discuss with him in person, perhaps for reasons of security, before sending the final casting instructions to Bersham where engine parts were being made and where he was still improving his cylinder-boring machine.

John was keenly conscious that these first two engines must succeed, and particularly his engine at New Willey. Each of them was put to a different job of work, the first one merely to pump water, but the Wilkinson engine is to be used in conjunction with his iron bellows to blow the furnace, a much more sophisticated task.

Here, in a nutshell, is the future engine market. First an engine that could pump water efficiently – out of flooded mines or from a lower to an upper pool to be used again by a waterwheel – but perhaps more important, an engine that could economically provide the motive power to drive all manner of machinery at an industrial site. John had seen the issues. Here was that tremendous confidence of success so typical of the man, and the more remarkable because he was poised right on the leading edge of the new technology. He recognised Watt as the key figure to the future but had seen his need of reassurance and support, which he was quick to provide, at the same time shrewdly trying to secure for himself the engine market for castings and cylinders.

That autumn the two men were constantly in contact over the preparatory details of drawings for valves and small engine parts to be made at Soho, and for the iron castings made under John's direct supervision at New Willey for his engine – while the Bloomfield Colliery engine castings were made by his brother at Bersham. John was impatient to get his engine up and then to modify any problems in practice; Boulton was urging care and caution to get it right first time, aware the world was watching. Watt stood uncomfortably in the middle, seeking to avoid the commercial stress in a full-time involvement with the engine.

By the end of January John was putting on the pressure. Watt had gone back to Soho to make some refinements and John was waiting impatiently at New Willey. There was an imperious note in the letter he now sent after Watt, perhaps as a consequence of this frustrating and unwonted dependence on another:

January 26th, 1776, Broseley.

Dear Sir,

Wm Thomas who is now over here from Bilston is ordered to send this forward imediately by William Johnson requesting the imediate dispatch of the needfull to put my engine at work. The loss I must sustain now every day will be considerable having provided men and stock coming in daily which cannot now be declined on an expectancy that we coud not fail of being at work before this. Simplifying the condenser in the manner you hint at will be a very considerable improvement in my opinion, but as I am circumstanced now I had rather have an engine at work on your old plan than suffer from longer delay.

I have requested Wm Thomas to procure a carriage to bring the articles wanted from you and Bilston directly hither. I hope we may have them with your company early in this next week. Wm Johnson can explain in part some of the hopes and disapointment that must arise in a furnace situated as mine is now at this place and whatever plan we get to work upon I beg it may be that which will be soonest done. Any improvements may be added at a future day when I can spare the time better ...[12]

On 5 April John was hurrying back to New Willey following a period at Bersham and was able to tell Watt that 'the engine goes very well', though they were still watching it carefully and making adjustments.[13] The Bloomfield Colliery engine was also pumping satisfactorily by this time and it was at this point that Watt took the opportunity to return to Glasgow to marry Ann McGregor and tidy up his affairs in Scotland. He was away most of the summer, during which time orders for engines were beginning to flow in. Boulton was impatient for his return in order to liaise with John and reassure the customers:

If we had a hundred wheels [i.e. rotary engines] ready made and a hundred small engines like Bow engine, and twenty large ones executed, we could readily dispose of them. Therefore let us make hay while the sun shines, and gather our barns full before the dark cloud of age lowers upon us, as to your absence say nothing about it. I will forgive it this time provided you promise me never to marry again.[14]

In 1776 Boulton was 48 years old, the same age as John but eight years older than Watt. He was perhaps feeling the onset of middle age before the others and his obvious anxiety that delays in executing orders would lead to lost business was to be prophetic.

Meantime, Watt returned to Birmingham with his bride and installed her in what can only be described as his bachelor quarters in Newhall Walk; a small town house dominated by servants, noisy with the two young children of his former marriage and cluttered with his papers and projects. It was not to her liking and within a few months the family had moved to an imposing three-storey house called Regent's Place at Harper's Hill, conveniently close to Soho.

John, too, was under family pressure at this time. It was some twelve years since his daughter Mary had returned to live with her father and stepmother at The Lawns, Broseley, and she had grown up into a confident, educated young woman with poise and presence. In the absence of anything more than brief references to her in the earlier letters it is yet important, in view of later events, to try to interpret her father's plans for her at this point.

He had no sons nor could expect any from his second wife, Mary Lee, in view of her age. All his great contemporaries, including Watt and Boulton, had sons. John's power and possessions were growing rapidly, but without a son to inherit how did he regard his daughter? It must be that he hoped she would marry someone with a good base and good connections in the iron-making industry, someone who would one day be worthy to inherit his empire. The education he had provided for Mary, and her closeness to him and his affairs during these important years, was perhaps part of a process that he hoped would help her to choose well.

Time after time during his frequent absences around the country, John left his wife Mary in charge at Headquarters, as he called his Broseley/New Willey iron-making base; an unusual and important responsibility for a woman of that age and particularly in an industry almost exclusively male. Daughter Mary would grow to womanhood within that system and would certainly have more experience in treating with men on a day-to-day basis than most young women of her time. There can be little doubt that John encouraged and lauded this independence of mind and spirit. Was it part of a long-term plan? If so, there is evidence it backfired.

By early 1776 John's daughter Mary, then aged 19, became engaged to a young and consumptive physician called Richard Blackley. How much parental opposition there had been to the proposed marriage is not known, but that it proceeded to a public betrothal probably means that they were resigned to it. So much was happening in John's business life at the time that it is possible he buried his personal emotions in his work, an escape mechanism he had used before, on the death of his first wife, and would use again. It is tempting, therefore, to think that the demise of the young doctor before the marriage took place would be regarded by John as a blessing. But the ensuing turmoil at The Lawns is not difficult to imagine either, compounded unhappily by the rapid decline and death that same summer of his nephew, young Dr Blakeway, the son of his staunch friend and business partner. The events would inevitably impact on his business life.

There is certainly a noticeable tetchiness in his exchanges with Watt and Boulton at that time, though it was not all down to domestic stress. He was still having trouble with his first engine whilst building a second at New Willey for pumping water back to the upper pool, and he was becoming increasingly frustrated at the delays in providing casting instructions and drawings for other orders which had been placed. This could only in part be attributed to Watt, since Boulton drew up the terms based upon the savings in fuel on new engine against old; a system made for disagreement and which led to continuing dispute, yet which had to be agreed before the final contract could go ahead. John was frequently held responsible for the delay in providing cylinders and castings, when in fact he was not a partner in the enterprise but more a sub-contractor. He increasingly tried to anticipate these final instructions and have the parts ready for when the authority to provide them came. It was an unsatisfactory and stressful arrangement, but the rewards were potentially great. The letters of the summer and autumn of 1776 illustrate the difficulties:[15]

28 April, John at Broseley to James Watt at Soho:
… I beg we may have drawings or some instructions for the other engines in hand without which we can't get forward. The little articles that come often at last retards more than castings …

30 April, John at Broseley to James Watt at Soho:
Yesterday morning about 6-o-clock we stopped and took out the regulators to see if we could make any alterations for the better, at same time examined the junction with the copper bottom and found your joint exceedingly defective, so much so that light may be perceived from a candle placed within the cylinder. You must in future adopt a different method in the joining this part of the steam pipe to the inner bottom as a joint there is not to be come at …

1 May, John at Broseley to Matthew Boulton at Soho:
… The cylinder & working barrels for Bedworth <u>are ready</u> [John's underlining]. The pipes are casting. Nothing shall wait if you only take care I

have the needful instructions in time. If I was a taylor I should be inclined to remark that it was more difficult to get the measure taken than to make the suit of cloths …

7 May, John at Broseley to Mr Watt at Soho:
I most heartily concur with you in the cast iron bottom recommended in your favour of the 4th. It would have been a very capital improvement in our present engine had it been adopted. Don't you think it will yet be best for me to have such a bottom in the room of the copper one here …

9 August, John to Matthew Boulton Esquire at Soho:
… I wrote you on Wednesday from Salop with Bill of Lading for sundries for Bow Engine. I now enclose particular charge of the whole. Had all the order been given at once, the whole would have been up together with the cylinder. In future I wish to have orders for all together. The small articles for which instructions have been last sent take up most time. But … a little more practice will put every department in a more expeditious mode …

12 November, John in London to Messrs Boulton & Watt:
Gentlemen … I have seen Mr More … The misfortunes at Bow afford great room for exultation among your enemies. Mr More has been told by several that the engine will never answer. I hope you will lose no time in putting this affair right as soon as may be …

John, and with some justification, felt himself to be a principal participant in the engine business and spoke freely and on equal terms to both Watt and Boulton. His engines were the exhibition engines on whose workings, as a considerable practical engineer in his own right, he fed back information and suggestions to Watt. It is clear that he saw them as working engines that must be good enough to promote further sales and, as such, should have instant attention if anything was wrong. His warning here, to attend urgently to the breakdown of the engine erected at a distillery at Stratford-le-Bow, show he was aware that all the early engines were important exhibits that must be seen to succeed, most particularly this one in London.

It was John, too, who arranged the first sale to Cornwall, an important established mining area with an urgent need of better pumping systems, and a large potential engine market they had been seeking to procure for some time. He also proposed to Boulton that, as it was such an important market, they should be prepared to take risks and go ahead quickly with this first engine for the West Wheal Virgin Mine without the usual contract securities. Boulton was cautious; John was impatient; Watt listened and went on producing the drawings. They did secure this market, and over the next few years each of them at different times made the long, hard journey to Cornwall for a protracted stay (which they did not enjoy) to promote and erect, and then to modify or repair, their engines. It must have been worth the trouble.

John at this point had sounded out Boulton and Watt about his entering into the partnership on a formal basis, and when the inevitable cash-flow problem attendant on new businesses overtook the partnership in these early years, with capital outlay not yet recouped by income, Watt wrote to Boulton from Cornwall where the biggest capital drain was occurring to remind him of this.

> You know I am a bad ways & means man, but however, the following thought may merit your consideration. You know Wn. [Wilkinson] has several times hinted a wish to be concerned in this scheme, to which we have had material objections but rather than founder at sea we had better run ashore.[16]

Not once, then, but several times John had made approaches to them about being formerly admitted to the partnership, but Boulton had resisted. Why? There was often a frisson of tension between these two men. Was Boulton wary of John's confidence and drive? Certainly this rejection will have smouldered with John to become another reason for the break-up of their association in the years to come.

In the summer of 1776 contact between John and his close friend Samuel More, now secretary of the Royal Society for Arts, Manufacture and Commerce, led to a visit by More to the area to inspect the various new developments on behalf of the society. More arrived in Birmingham about midday on 10 July and John joined him at his inn during the afternoon. They looked at the canal but did not go to Soho, and both slept at the inn that night. It is clear that, as two close friends, they needed uninterrupted time together to talk through the recent events and plan the visit.

The following day, accompanied by Josiah Wedgwood, they were Boulton's guests at Soho. They dined there and Boulton showed them the memorial garden to Dr Small he had laid out in the grounds. He recounted, for More's benefit, the history of Watt's engine and they examined the old Kinneil engine now set up at Soho with a Wilkinson cylinder. Then an astonishing thing happened. The four men took off their coats, rolled up their sleeves and dismantled the engine piece by piece to understand the detail of its working. They then reassembled it and had it running satisfactorily again by evening. Watt was not there. They spent that night at Soho.

The next morning, 12 July, John and More took their leave and went to John's Bradley works to show More a 'fire engine' set up there to blow the furnace through iron bellows. It was the forerunner of John's New Willey installation; More saw it working and was impressed by the blast generated. The following day he left John to inspect the Bloomfield Colliery pumping engine. He was engaged to meet Boulton and Wedgwood there but they failed to appear on time, so he persuaded the engineer, a Mr Perrins, to show him round and noted his approving remarks. Boulton and Wedgwood eventually caught up and Boulton took him in the afternoon to the quarterly meeting of ironmasters at Stourbridge, where trade values and prices were to be agreed for the three months ahead.

For the whole of the next week, More was John's guest at The Lawns, where everyone was in distress at young Dr Blakeway's rapid decline. It took them about four hours to reach Broseley from Birmingham, travelling in a chaise and changing horses at Bridgnorth, and they lost no time in escaping from the unhappy household to see the works at New Willey in full production. Over the next few days, More, with John as his guide and mentor, examined every aspect of the New Willey Works and his journal provides important details of the installations. He also visited John's lime pits nearby and saw his cast-iron 'rail-roads', which linked the works to the Severn. They discussed Wedgwood's idea for 10 miles of double track cast-iron rails from Derbyshire limestone quarries to his house, costing £1,000 per mile. During his stay at Broseley, More also crossed the river on the ferry to visit the Darby works in Coalbrookdale and the Reynolds works at Horsehay and Ketley. He dined with the Reynolds and met young William whose laboratory and collection of ores impressed him.

It was an extremely busy itinerary; a build-up by John for what was to come on their last day together, when he took More to see the model of the Iron Bridge. Unfortunately, Abraham Darby was not at home and they were unable to view the model itself. The day was instead spent looking at the site for the bridge and discussing the need for a better, quicker link between the different works on opposite sides of the river. John's advocacy of the bridge was clear in More's journal.

By using cast iron, everything could be made ready with minimum interruption to the navigation when the work started. There was to be one arch, 40ft high, to further facilitate the passage of masted vessels, with a roadway 16ft wide across 120ft of river. This bridge, they believed, would lead to other bridges which would further benefit the area's important works. If no one could tender for less than £2,000 'Mr Wilkinson has engaged to complete it for that sum'.[17]

Here is that confidence again, and an enthusiasm for the project that comes through the dry words of the journal. The next day, More left Broseley for a visit to Wedgwood's works at Etruria before returning to London, and John set him on his way as far as Shrewsbury. They clearly still had much to talk through before John returned to his sorrowing household and to another significant threshold in his life.

Following the visit of Marchant de la Houliere the previous year, William Wilkinson was to go to France with all the implications for the continuing management of the Bersham Furnace. He was still in charge at Bersham in early May, consulting with his brother at Broseley before casting a cylinder bottom of a new design.[18] He was at Broseley for part of Samuel More's visit, too.[19] But by early September he had gone and John was by then constantly on the move between his three main works at New Willey (Broseley), Bradley (Bilston) and Bersham; actually supervising some of the work at Bersham himself early in September and suffering 'a burn caught at the casting of a brass plate, which is likely to confine me some time'.[20]

His sheer appetite for work and new business was staggering. With William in France, and in a good position to secure more orders there, John was yet prepared to add the Bersham work to his burthen. Though perhaps he was not unhappy at again securing close control of that large, iron-making establishment at a time when orders for iron castings, as well as guns, were climbing steadily. Which meant, then, the three large Wilkinson works were under his direct control.

With all this happening in his business life a quiet period of consolidation might have been expected. Nothing of the sort! Soon after William left for France, John became embroiled in the developing politics and manoeuvrings of the project he had enthused about to More and the one closest to his heart. The great Iron Bridge over the Severn was about to be built.

6

THE IRON BRIDGE

By the early 1770s John had three large works producing iron. From the position of authority and influence this gave him among the Midland ironmasters, he missed no opportunity to promote iron and enthuse about it.[1] His detractors called it his 'iron madness'. It is hardly surprising, therefore, when the increase in industrial and commercial activity between opposite sides of the Severn Gorge in the Coalbrookdale area demanded better communications in the form of a bridge, John should press for it to be built of iron.

The summer of 1774 saw discussions gaining momentum between the ironmasters and those who owned land on either side of the river, or who had regular need of the existing ferry, sufficient to produce reports in the West Midlands press at the time. A first formal meeting took place on 15 September 1775 in Broseley, where John lived. From this point forward the chronicle of events leading to the building of the bridge and its early period of operation was laid out, step by step, in an insignificant little red minute book (now in the keeping of the Shropshire Archives), in which a certain Thomas Addenbrooke recorded every detail in a clear careful hand.[2]

John was involved from the beginning, and though some commentators have been deceived into thinking that a conventional bridge was first planned – by the use of words like 'intended bridge' or 'proposed bridge' and by the absence of the word 'iron' to describe the bridge until well into the chronicle – there is evidence that this was not so. As early as 1773 Thomas Farnolls Pritchard, the subsequent architect for the Iron Bridge and by then John's friend of some fifteen years standing, discussed in a letter to him the prospect of an iron bridge in the Severn Gorge.[3] There is the draft of a petition to the House of Commons, for obtaining the Act of Parliament to build the bridge, handwritten in the very back of the little red minute book:

> To the Honourable the Commons of Great Britain in Parliament assembled the honourable Petition of the Gentlemen Clergy Merchants Manufacturers and Tradesmen residing in or near Coalbrookdale Madeley Wood Benthall and Broseley in the County of Salop whose names all hereunto subscribe SHOWETH that a very considerable traffic is carried on at Coalbrookdale

Madeley Wood Benthall Broseley and places adjacent in iron lime pottery clay and coals and that the persons carrying on the same are frequently put to great inconveniences delays and obstructions by reason of the insufficiency of the present ferry over the River Severn ... particularly in the winter season in which time it is frequently dangerous and sometimes impassable ... the trade would be much improved as also the communication between the several places beforementioned ... if a bridge was erected across the said river ... the said bridge to be constructed with cast iron ...[4]

No date is affixed to this draft, but the Act of Parliament was on the statute before 15 May 1776, which, given the time required for its introduction and the various readings in the Commons, puts a date on the draft of early 1776 and possibly late 1775.[5] In fact, at the very first recorded meeting of the subscribers a minute is passed to appoint a lawyer to obtain the Act of Parliament. The draft

Thomas Farnolls Pritchard, architect of features at The Lawns, John Wilkinson's 'Headquarters' in Broseley, and of the Iron Bridge.

is likely to be the consequence of this meeting, and leaves little room for doubt that, from the very beginning, the subscribers were planning to build in cast iron. Abraham Darby III was 'chose treasurer' though not present at the meeting with Pritchard commissioned to prepare estimates of the cost of the bridge; the former the head of the large Darby iron-making complex in Coalbrookdale, the latter with a stated interest already in building in iron.

Pritchard lost no time in preparing the necessary details. An architect's drawing entitled 'Design for a Cast Iron Bridge between Madeley and Broseley' and carrying the details 'F Pritchard. Salop Oct 1775' is reproduced in an 1832 publication by a man called John White.[6] The original drawing has been lost but was obviously available to White and the details tie in closely with an important minute in the little red book, dated 17 October 1775:

> Agreed that the sum of three thousand one hundred and fifty pounds already subscribed be paid to Mr Abraham Darby, he the said Mr Abraham Darby in consideration thereof to defray all expenses in erecting the intended bridge in a substantial manner according to a plan this day produced by Mr Thomas Farnolls Pritchard or as near it as may be for the best and safest manner for making roads at both ends of the said bridge and obtaining the Act of Parliament for the same, he the said Mr Darby having any and all further sum or sums of money that shall be given towards building the said bridge or obtaining the Act over and above the said sum of £3,150 but in case there shall be any opposition to the obtaining of the Act then the said Subscribers agree to defray the additional expense ...[7]

It is worth examining how the money collected from the subscribers was gathered up. At the second formal meeting, on 28 September 1775, a list of seventeen names, destined to be closely involved with the future planning and management of the bridge and with John Wilkinson and two friends at the head, subscribed the relatively small amount of £23 5s 11d on a call of 10 per cent. A further sum of £18 6d total was later additionally collected from four of these names and one new name. On the next page in the minute book is another list, which comprises nine names already listed plus five new names, including John's landlord at New Willey, Squire Forester, and Sir Harry Bridgeman, who subscribed £50 each. John matched their contributions with another £50, making him the principal subscriber in the total of £283 1s 11d collected and recorded so far.

This looks like the necessary administrative fund to get the project started. It is a long way short of the £3,150 handed over to Abraham Darby III less than three weeks later, which will have been collected from the subscribers on the

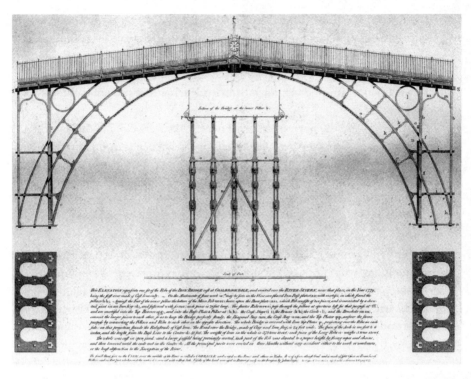

Design for the Iron Bridge.

basis of their shareholdings – sixty shares total at £50 a share. It might have been supplemented from an impressive list of more than fifty names, headed by Lord Craven, Earl Gower and Lord Pigot, who were added to a minute agreeing 'that the following noblemen, gentlemen, etc, be named as Commissioners'.[8] They would of course have given the new project prestige and status outside of any financial contributions they might have made.

Two questions emerged now that a substantial sum of money was available and a group of subscribers formally identified to carry the work forward. If John, with Pritchard, conceived the revolutionary idea of an iron bridge in the first place, why did he not at this stage undertake the iron work of the bridge himself? And was Abraham Darby III, a comparative youngster, capable and confident enough to cast the huge spans that Pritchard's design demanded even with all the experience of the Coalbrookdale company behind him?

John's position is not difficult to explain. Concurrent with the bridge planning was the introduction of Watt's new steam engine and the rapidly expanding demand for iron cylinders and castings from John's works, in addition to steady government orders for his guns. He was at full stretch to supply existing demand for iron products, on top of which his brother was about to leave Bersham for France, increasing his supervisory and management responsibilities. John was shrewd enough to see that compromise was required, that his influence must be behind an iron bridge, but that someone else would build it.

The Darbys were the obvious choice. Their works were close to the proposed site for the bridge and they had generations of experience as ironmasters. There was also a tenuous link with them going back more than twenty years to Isaac's arrival in the district and his lease of a Darby furnace. There might even have been an element of patronage in John's approach to young Abraham Darby III, though exactly how Darby came to be proposed and chosen as the bridge builder is not recorded.

From the beginning, there was evidence of an anxiety in the Darby camp that the money would not be enough and that they were engaging in frontier technology, the cost of which could not be precisely forecast. Abraham's fears were not articulated in so many words in the minutes, but the subscribers were at great trouble to reassure him. Yes, they conceded that if there was opposition to the Act there would be more expense, and that they would defray that cost. And yes, certainly, any further subscriptions received should be paid to the Darbys.[9]

How far these reassurances were successful has to be in doubt as a consequence of an entry in the minute book three months later, which was surprising and irregular. A meeting was recorded as taking place on 22 January 1776 at Abraham Cannadine's house in Broseley, which was the subscribers' meeting place:

> It was agreed that a new subscription paper be prepared with a new preamble setting forth that the several subscribers advance a proportion of their money subscribed towards obtaining an Act of Parliament for building a bridge from Benthall to Madeley Wood at the place first agreed. It was likewise agreed that no opposition or intention of an opposition was or is intended to be made to the erecting a bridge over the Severn at the Sheepwash by any of the proprietors of the intended bridge from Benthall to Madeley Wood.[10]

No list of persons present was given, the minute was not signed and there were no other minutes. Moreover, the handwriting of the first long sentence looks suspiciously like John's flowing, well-formed hand. The handwriting of the remainder is clearly by someone else.

What was happening here? A new subscription paper was called for, allocating only a proportion of the money already subscribed to the original bridge, and another bridge was proposed lower down the river, to which 'no opposition or intention of an opposition was or is intended', presumably by the subscribers. It can only have been designed to put pressure on someone, but whom? And who exactly was applying the pressure? Were the subscribers now split?

The next two meetings make things a little clearer and provide further evidence that Abraham Darby III continued to be worried about costs. On 25 April 1776 they discussed ways of limiting the competition to the bridge in the form of boats and ferries. And in the important meeting of 15 May 1776 a further charge on the proprietors was agreed to repair approach roads – though this had been specifically included in Abraham's overall responsibility when he was given the money more than six months previously. That was not all, however. Two further minutes come as a shock:

It was agreed to rescind minutes entered into with Mr Darby for erecting an Iron bridge over the Severn between Benthall and Madeley the 17th October 1775 …

… Agreed that an advertisement be inserted in the next Birmingham and Shrewsbury newspapers to be continued three times, that any person willing to undertake the building of the intended bridge from Benthall to Madeley Wood, one arch 120 feet span, the superstructure 18 feet in the clear and the centre 35 feet above low water, the proposals to be sent to Thomas Addenbrooke before the 20th June next.[11]

Abraham Darby III clearly wished to be free of his commitment to build an iron bridge, and the committee agreed to release him. At the same time they choose to go ahead with a bridge as planned and to the original specifications. But, since iron is not mentioned, they seem prepared to consider a conventional design. There was a further development at the next meeting on 28 July 1776. Presumably no satisfactory tenders had been received by the deadline of 20 June and there had been further discussions resulting in a modified plan for the bridge, again produced by Pritchard.[12] This plan has not survived. How far it differed from his original is not known, but the committee now agreed 'that advertisements be inserted in the Birmingham and Shrewsbury newspapers for persons to undertake the stone and brickwork'.[13]

There was now a ten-week gap in the record. It would be invaluable to know how the decisions and dealings of the committee up to this point affected the relationship between John and Abraham Darby III. They were, so far, the only ironmasters involved with the resources to erect an iron bridge. It is also important to remember that Abraham Darby III was only 25 years old; not yet fully come into his powers; not yet confident, perhaps, to treat with a man like John, 48 years old, a shrewd businessman and a very experienced ironmaster. Indeed it may be that the later appearance of Richard Reynolds on the committee of subscribers was a studied move on the Darby side to support young Abraham. Reynolds had earlier been totally responsible for the Coalbrookdale company and was a highly respected figure in the area. He was also an ironmaster with business knowledge and experience to match John's.

It is not difficult to imagine the frustrations surrounding John at this time: urgent in his enthusiasm to see the iron bridge built; unable to commit more of his energies and resources to it; determined to see the project through and casting around for a way to get it started. From what happened next it seems clear that he found a formula to persuade, perhaps even to compel, Abraham Darby III and the Coalbrookdale company to undertake the work and to get on with it quickly. The next meeting after the ten-week delay has this minute:

> Mr Jennings and Mr Wilkinson agree to let Mr Darby have their shares in consideration of his giving them security that an iron bridge shall be erected from Madeley Wood to Benthall in two years from Christmas next and Mr Darby agrees to take the said shares on that condition.[14]

Leonard Jennings, a local miller and a wealthy business man, was the third name in the original list of subscribers; coming immediately after John Wilkinson and Edward Blakeway. Blakeway was absent from this meeting and so retained his shares. Who knows what behind-the-scenes manoeuvrings prepared the ground for this agreement? It cost John and Jennings their shares in the scheme and gave Abraham Darby III control of the whole enterprise with thirty-seven of the sixty shares. But it also required him to undertake the building of the bridge, to build in iron and to complete it in the next two years.

The details were compellingly tied down at the next meeting, less than three weeks later, in a manner typical of John's business dealings elsewhere. First, provision was made for those lesser subscribers on the committee who were anxious about escalating costs, by obtaining from Abraham Darby an indemnity for them against any further charges on their shares. The subscribers concerned were identified in the minute book and were given further reassurance by a security from Darby to pay 5 per cent per annum on their shares from the tolls once the bridge was built. Their full voting rights as subscribers remained unaffected.

The need to raise any further money was then limited to a call on the remaining shareholders. Abraham Darby III and friends held thirty-seven shares, John's remaining friends held only five, which meant that if further money was required to complete the bridge Abraham and friends would have to find most of it themselves. The whole arrangement was vintage Wilkinson, but there was more. The most important detail in these minutes concerned the building of the bridge itself. It was couched in precise legal language and tied down Abraham Darby III very closely:

> ... the said Mr Abraham Darby doth agree to erect an iron bridge in one arch 120 feet span and the superstructure not less than 18 feet wide from Benthall to Madeley Wood to be completely finished with roads and avenues leading to and from the same as described in the Act of Parliament on or before the 25th day of December one thousand seven hundred and seventy-eight ...[15]

The meeting closed with agreement that proper assignments of the share allocations, as specified in the minutes, should be made the next time they met. Having surrendered his shares, John could not now vote at this meeting, but there at the end of the minutes, alongside the signature of 'Thomas Addenbrooke, Secretary', was his own bold signature as 'Witness'.

Nothing now happened until 31 March 1777, almost six months later, though two earlier meetings had been postponed which might have been evidence of new manoeuvrings. In the interim someone, who is not identified, questioned the legal validity of agreements and share allocations drawn up at the meeting of 18 October 1776. It is unlikely to have come from the Darby camp, in view of remarks in a letter written in November 1776 from one of the Darbys to a relative in Sunderland, which suggested Abraham Darby had taken his commitments in that October meeting seriously and was already making plans:

> ... The Bridge, that is to be made over the Severn at the bottom of the Dale is now fix'd upon to be an Iron one, wch. certainly when completed will be one of the great curiosities yt this Nation or any other can boast of ... I suppose it will all be cast in the Dale for Cousin Abram. will have the whole direction ...[16]

There is an evident pride in these words: no sign here of a young man seeking an escape from an onerous undertaking; indeed, the reverse.

If, then, John initiated the questions, did he seek a new share allocation in which he could again participate? Did he and Jennings surrender their shares to Darby knowing that this would be possible? The questions might have come from the nervous group of lesser subscribers who sought to have the promised 5 per cent per annum on their shares absolutely guaranteed by legal opinion. If so, they were to be disappointed.

Thomas Addenbrooke was authorised by the committee to attend on Mr Thomas Mitton of Cleobury so that the agreements and allocations of the earlier meeting could be drawn up in proper legal form to be signed and made binding.[17] There were unspecified difficulties which he brought back for discussion by the committee, and which led to the following minute:

> Mr Mitton's opinion on the cases referred to him at the last meeting were produced but not being quite satisfactory to some of the Proprietors present, Mr S Roden, Mr Blakeway, Mr Morris and the Secretary are desired to draw up a case in such manner as appears most eligible to themselves and bring his opinion to our next meeting.[18]

Mr Mitton had made things very clear. No security to pay interest on shares could be given under the Act. The only way interest could be paid to shareholders was in the form of a dividend of the tolls from the completed bridge and approach roads. The whole minute of 18 October 1776, in which the new agreements and allocations had been made, should be rescinded.

This took the basis of the shareholding back to the sixty shares originally subscribed and paid for with a proviso that four more shares should be raised. Legal assignments for this disposition of shares were to be drawn up immediately, which of course brought John and Leonard Jennings back into the frame as potential shareholders. Had John foreseen this? It is likely he would have gone to great trouble to find out what was possible, and what was not, under the Act. It is also probable that he would have known, and that young Darby in all likelihood would not know, that what appeared to be a total surrendering of Wilkinson shares and power in exchange for the undertaking to build the bridge in iron, was in fact a cunning short-term device, which would serve his purpose at the time to start things moving, but which could not be supported in the long term by law.

If this were so, there will have been one big question troubling him now. Did this new reversal mean that the agreement for Abraham Darby to build in iron to a time limit, established at that same meeting on 18 October 1776, was as invalid as the attempt to guarantee an annual percentage payment on the subscriptions? John may have argued that the legal wording, the whole minute, referred to the shareholdings and proposed interest payments alone, which is what Mr Mitton had been asked to pronounce on, and not to the bridge-building contract with Abraham Darby III. However, it is also probable that Abraham Darby would not wish to be seen to procrastinate further once his financial security could be reasonably established, and the rest of this important meeting was given over to the terms under which he was to proceed.[19]

He had to build in iron, though now to a slightly altered specification. The span of the arch was reduced to 90ft from 120ft, and the width of the superstructure increased to 24ft from 18ft. There was to be a proper tow path under the bridge, and Darby was responsible for all roads leading to the bridge at either side and for all toll gates and turnpike houses required. He was still tied to the same time schedule, even though nine months of the twenty-six had elapsed since he first agreed it. Moreover a penalty was now attached under which he was to pay 5 per cent on all monies paid to him from the time he received them, if he failed to complete the bridge by Christmas Day of the year following.

Provided these terms were fulfilled he could retain the money arising from the shares including the four additional shares. A further 5 per cent was to be raised immediately from forty-six of the shares and 2.5 per cent from the remaining eighteen shares (the nervous subscribers). But then came a generous additional clause: another 10 per cent on all shareholdings was to be paid to Abraham Darby as soon as the bridge abutments were finished, with an additional 10 per cent every three months thereafter until the bridge was complete. In addition, all monies received from the toll gate and the ferry before the bridge was ready for use were to go to him.

All the members present at this meeting had signed the minute book in their own hand. It was the first time this had been done and obviously marked a significant occasion. The committee had gone a long way towards removing Abraham Darby's financial concerns; by allowing him to retain all the money so far received and by providing a further small income flow if that should prove necessary. At the same time, they re-established the pressure on him to complete the bridge and associated works in the strict manner proposed and to a strict time limit, with penalties if the time limit was not met. Why such a rigid contract was considered necessary is not made clear.

The purpose of the next meeting was to formalise the agreed business of the last.[20] Sixty-four share assignments were signed and sealed, though Thomas Addenbrooke lists only sixty-one of them in the minutes.[21] Of these, John had twelve and Leonard Jennings ten, with Edward Blakeway and Pritchard taking two each. Abraham Darby himself had fifteen shares and one of the four new shares went to his brother, Samuel. The only other substantial shareholder was the Reverend Mr Edward Harries who owned the land on the Benthall side of the river, the remaining shares in ones and twos belonging to the nervous subscribers.

Thomas Farnolls Pritchard died about this time and never saw the completion of his design; his shareholding went to his brother. John and his friends were again the dominant shareholders and remained in a powerful position to influence decisions about the bridge. Shortly after this, however, John's name disappeared from the minute book and he did not attend the meetings of trustees for the next four and a half years. His brother-in-law, Edward Blakeway, continued to be involved and as he lived in Broseley Hall, just across the road from The Lawns, he would obviously have kept John closely informed of developments throughout those years, which were going to be some of the busiest of John's life.

During this period the Iron Bridge was completed by Abraham Darby III, and to enormous acclaim. It was not completed by the deadline of Christmas Day 1778, though no penalties seem to have been exacted from him on this account, which perhaps reflected the satisfaction, even admiration, of the subscribers, including John, for the manner in which he tackled and completed the work. There was, however, an arrangement in the minutes for providing the nervous shareholders with extra money for road building and repairs, in part from Abraham Darby III and in part from the road tolls, and perhaps this was an alternative to penalties acceptable to all.[22]

Landscape painting of the finished Iron Bridge.

The recent discovery in a museum in Stockholm of a painting by Elias Martin, an eighteenth-century Swedish Professor of Art, has thrown new light on theories as to precisely how Abraham Darby III raised into position the huge castings which supported the bridge, about which process no written or diagrammatic records have survived. This painting shows the bridge under construction with three of the great spans in position, secured there by simple scaffolding and apparently raised into that position by no more than wooden ships' derricks. With a system of pulleys, which along with simple derricks would be the essential tackle of any eighteenth-century naval or merchant shipyard, the only ingredient lacking would be manpower. Men, lots of them, hauling on ropes controlled by other men with established stop/start signals; slowly and carefully moving into position high above the river those huge spans of iron; the spans finally being fixed in place by a few who took their lives in their hands to move nimbly over the spare wooden scaffolding to secure them. A false move and the castings, of a size never seen before in Coalbrookdale or anywhere else in England, could have slipped and crashed down destroying scaffolding, derricks and the gangs of men below. Small wonder that there was evident pride in the achievement, which was shared by workmen and subscribers alike. Few in the area would be untouched by it. No subscriber, therefore, would wish to exact a contractual penalty from Darby for not completing the project on time.

Further evidence to support this method of constructing the Iron Bridge has now been supplied by the erection of a half-size replica of the bridge a few miles further down the Severn Gorge, constructed using no more than simple eighteenth-century technology and manpower, supplied by the army, under the direction of an engineer commissioned by the Ironbridge Gorge Museum Trust Ltd. During the successful undertaking of this project, which was filmed by the BBC *Timewatch* programme, the research team were impressed by the understanding of applied physics and geometry required by Darby in the various stages of construction, and were able to understand why it was necessary to have available the large quantities of rope and timber listed in his account books and inventories.

From the time of the legal assignation of the shares in the autumn of 1777, Thomas Addenbrooke's recordings of meetings of shareholders in the minute book refers without exception to the 'intended bridge' until 22 July 1779, when it becomes the 'iron bridge'. This indicates that their bridge was now a fact and was in position, though the bridge roadway and the linking roads were not yet complete. Twice during the next fifteen months there were calls on the shareholders for small additional sums of money as per their agreement with Abraham Darby if there were delays in completing the bridge. It was finally opened to traffic on New Year's Day 1781, at a stated cost to that date of £2,737 4s 4d. Since the toll house and other gates and roads had yet to be completed, the figure was probably for the bridge and abutments only.

The pride and independent spirit of men like the Quaker Darbys, indeed of all the shareholders, is evident in the wording at the bottom of the table of tolls mounted on the bridge soon afterwards: 'N.B. This Bridge being private property, every Officer or Soldier, whether on duty or not, is liable to pay toll for passing over, as well as any baggage wagon, Mail-coach or the Royal Family.'

It would have been particularly satisfying to John to see those words posted, for he firmly believed that manufactures and commerce 'will always flourish most where Church and King interfere least'.[23]

The business conducted by the trustees at their subsequent meetings gives vivid glimpses into the daily life of late eighteenth-century England, most of them concerning issues still with us today: bills were posted offering rewards for information about persons evading the tolls; 'for the discovery of the person or persons who broke the two ballustrades of the cast iron bridge in the night'; for 'the Constables who summoned the persons for evading the tolls … to be allowed two shillings and sixpence each day they attend the Justices'; and for George Armstrong's hogsty to 'be removed to a proper place'.

More events played out on the bridge. The proprietors of the fast coach between Shrewsbury and London, *The Diligence*, negotiated a special rate with the trustees to be paid quarterly in arrears and then had to be pursued for payment. Messrs Banks and Onions continued to allow the spillage of 'cinders, ashes and other obstructions' to pile up beside their furnace on the approach roadway on the slopes on the Benthall side and had to be cautioned by the trustees. There were difficulties about taking down a popular brew house which the trustees eventually agree to pay to re-erect 'in a place that will not incommode the road leading to the bridge'. They then had to ask the owner to 'remove his pig and pig troughs out of the road'. Mr Bishop, the Distributor of Stamps, went directly to the commissioners to inform on the gatekeepers at the bridge who had been taking tickets from post chaise drivers – the trustees wanted to know what he had said, and why he had gone over their heads.

Meantime, John had moved on. With the bridge a triumph in iron, albeit for the Darbys, with steam engine sales steadily expanding, though with little hope for him of a real partnership with Boulton and Watt, and with promising signs of new business following his brother's successes in France, he retreated again to the northern haunts of his youth, to Wilson House and Castle Head, where new plans were afoot.

7

THE NORTHERN SANCTUARY

Since Isaac Wilkinson retained the property at Wilson House near Grange-over-Sands when he moved to Bersham in 1753, it is likely he intended to maintain contact with the people and the places of his northern beginnings, possibly for business reasons. Good quality iron ore was available in Furness with the coastal route south to the Dee ports providing convenient access to his new ironworks at Bersham. Isaac's focus changed to South Wales, however, and there is no record of his ever returning to the north.

Not so with John. In the twenty years following his arrival in Bersham in 1756 it is clear he retained a bond with the place of his youth. There are a number of documented returns with property purchases around Wilson House, which were perhaps intended initially to do little more than extend the holding there.[1] But in the summer of 1778, with the building of the Iron Bridge by then assured, other plans were forming. He returned to Wilson House for a stay of five months and set up an experimental iron-smelting base there using one of the early Watt steam engines. His letters to James Watt at that time are revealing:

> … Getting sufficient steam with peat was what I wanted most. That suspicion is totally removed, for I observe we can work with any rubbish whatever. The making of iron with this sort of fuel will be the next concern, and of which I hope to be able to give you some account in a fortnight or three weeks … If our regulating beam had been longer I fancy we should have done better with our mechanism as to opening and shutting the regulators. We are too near the stroke with our plug. We had a good deal of trouble with the perpendicular hanging valve at the hot water pump and was obliged to alter it before we could get it to draw sufficient hot water … Mrs Wilkinson and my daughter are with me. They desire their best respects to you and Mrs Watt …[2]

John used the opportunity arising out of the setting up of another new Watt steam engine to observe and comment on the progress and difficulties he experienced. With James Watt at the time on a protracted stay in Cornwall doing precisely the same thing with their Cornish engines, the exchanges between the two men at this time carry additional point. It seems likely, too, that with Watt absent for a long period in Cornwall John had decided it was a good time to be away from Birmingham to further his own plans in the north. Yet the two men maintain a steady, if slow, exchange of information by letter. John wrote to Watt again:

> ... We began to blow again here on the 25th past with half charpeat half charcoal and make good strong metal though it comes <u>very dear</u> [John's underlining]. This day all peat coal is put on. Next trial will be half peat coal half raw peat, then charcoal and raw peat ringing all the changes I can think of [to] procure a metal strong as bar iron if possible. The cost I shall not mind if I do but succeed in my pursuit of the strongest cast metal, which I flatter myself I shall find out tho' as I observed before it will be very dear.
>
> Your company for a week here at this time would be very pleasing. The different fluxes – and cinders – together with the different metal would I think be a high treat to you – exclusive of the engine which goes very pleasantly with any kind of rubbish peat or even peat mull. Thomas has been proposing today to try whins and savin, from which you may infer that we can manage any sort of light fuel – and that this engine may pave the way more readily for erections where coal is not to be had ... I am satisfied one of these engines may be worked with heath if no other fuel offered. A new boiler might also be constructed to answer light fuel still better – tho' this exceeds my expectations ...
>
> If Mr Boulton is with you I desire my best compliments, and here give me leave to congratulate you both on the prospect of your harvest in Cornwall. That is the country which will produce the best crop and is most worth your attention. Mrs Wilkinson and my daughter desire to be remembered to you all ...[3]

The timing of these experiments to produce 'the strongest cast metal' is concurrent with Abraham Darby III's trials to make the castings for the Iron Bridge of a size and weight that had never been attempted before. It would be fascinating to know whether or not there was any liaison between the two men at this time, or, conversely, whether John's removal of his experimental base to this northern outpost was planned to make it easier for him to keep any new discoveries to himself. That seems unlikely in view of his letters to Watt, though there is no documented information on a co-operation with Darby either, and absolutely no reference to him or the bridge in these letters. They do, however, provide useful information on John.

His wife and daughter accompany him on this extended stay in the north, his wife 55 years old, daughter Mary 23. Both are clearly familiars in the Watt and Boulton households and attach their greetings to his in the letters. For the time being, Headquarters had been moved temporarily from the comforts of Broseley to Wilson House, but this was also an indication of how important the two women had become to him. Was he also measuring their responses to this place which he was so drawn to himself? Were they part of another plan?

He was of course making himself tremendously busy with his experimental work at this time, a strategy which he had adopted at other stressful moments in his life and which he would use again. And he was missing James Watt, in whom he recognised a kindred spirit; an innovator and thinker who could complement his own experimental approach and enjoy the newness and excitement of what he was doing – regardless of their different values and ideas as human beings on other issues.

The placing of this distance between himself and the Midlands iron-making world at this time, and for so long, is itself significant. The great successes were falling to others: Boulton and Watt, Abraham Darby, even to young brother William in France. The years of growth and consolidation had made John a wealthy and acclaimed ironmaster in his own right. Yet were his achievements of a lesser order, destined only to feed the greater successes of others? Was he disillusioned? Did he need time to reflect, as well as to experiment? And was it out of this that his next great project emerged?

On 19 August 1778, as this important period of engine and furnace experimentation was beginning, John signed an agreement with six yeomen of Lindale which gave him control of a large tract of salt marsh between Wilson House and the sea, lying close under Castle Head Hill. It was intertidal land known locally as the Lindale Pool on which the six yeomen had grazing rights at low tide, but which was inundated at high spring tides along with some of their low-lying permanent grazing close to the marsh. John clearly wished to control the land between Wilson House and Castle Head and around Castle Head Hill itself. He had appraised the situation very shrewdly and the agreement was masterly: everyone gained an advantage; no money changed hands.

19th August 1778. By a Paper Writing or Memorandum dated this day. After reciting that they whose names were thereunto subscribed, being owners of the Lands and Meadows adjoining to Lindal pool had from time to time suffered great damage by the Tides running up the pool and overflowing their lands and meadows AND that Mr. John Wilkinson Iron Master had proposed to make a Bank from Castlehead across the pool to Low Meathop Land at his own expense to stop the Tides flowing up the said pool any longer UPON CONDITION that they gave up their right to him in the pool from Castlehead to Wilson House –

They taking the circumstances into consideration Did thereby for themselves and their Heirs give and grant to the said John Wilkinson All their right in the said pool or in the Sand-Land or Ground thereof so that the said John Wilkinson and his Heirs might enjoy the same for ever thereafter from side to side within the then Banks of the said pool without molestation from them or any of them – PROVIDED that the said John Wilkinson should the then next Spring make the said Bank across the said pool to stop the Tide from flowing up and should repair and support it for the future at his own expence –

Signed by Thos. Settle, William Turner, George Carter, Thos. Ryding, Edward Herbert & Thos. Hodgson.[4]

All those qualities of character associated with John's business life in the Midlands are here again demonstrated: shrewd appraisal of the situation; a nice balance of advantage to either side; imaginative foresight; and enormous boldness and confidence.

There are 10-metre tides at equinoctial springs in Morecambe Bay, and the Lindale Pool was open to prevailing south-westerly gales. A combination of the two produces a powerful and destructive sea and John was going to hold this back in the mouth of the Winster Valley on a 1-mile front to the east of Castle Head Hill, and across half a mile to the west. Moreover, for the agreement to stand he had committed himself to completing the bank during the six months of the winter, between the big autumn and spring tides – a bank, too, which would have to include some kind of bridge with tidal doors to open and let through the flow of the River Winster at low water but which would be pushed and closed by the rising tide.

He would not have undertaken this enormous task lightly. There is no direct evidence that he studied the building of sea banks either around Morecambe Bay or elsewhere in the country, but he had a close knowledge of this coast and the advantage of observing over many years the changing patterns of sand and silt in tidal scour and build-up in this corner of Morecambe Bay. A huge workforce would have to be assembled of men with horses and carts to move the sheer volume of material required for the bank, but John would have established by then that it would not have to be carted far. The sand and silt banks of the marsh itself would have provided much of it, stiffened by pinnel, the local name for the glacial debris of gravel and cobbles bedded in sand or clay, and lying in easily accessible banks all along this shore.

Despite this, it was an epic undertaking to be completed in the six months of winter, in a place which by its very location is exposed at this time of year to raw, wet, bitter winds that drove man and beast to seek shelter. John would have needed to see the work well underway before he moved south again just before Christmas.

Only after this agreement with the six yeomen was signed and he was sure the bank would be built in time did he secure the land where he had now decided to build, for himself and his family, a house of some style and pretension. In a deed dated 2 December 1778 he purchased from William Turner, one of the men in the Lindale pool transaction, Castle Head Hill. Described as 'all that parcel of woody and waste ground lying at or near Lindale called and commonly known by the name of Castle Head … and amounting to six acres or therabouts', the sale included 'the two closes of meadow ground amounting to two acres … situated at the west side of Castle Head called Castle Head Meadows'.[5]

How far this purchase depended upon the completion of the Lindale Pool transaction is a matter of speculation, though John did pay William Turner the then substantial sum of £350 for this land. He also took on responsibility for a number of annual payments attached to the freehold, like the small stipend to be paid yearly to the curate of Lindale. Four of the six yeomen of Lindale sold land around Castle Head to John in subsequent years, providing evidence that his side of the agreement had been kept and the bank had been completed on time. It served its purpose and kept out the sea for almost eighty years. Parts of it can still be seen in the valley to this day, though in 1856 it was superseded as a sea wall by the causeway and embankment for the new railway at the very mouth of the Winster Valley.

John decided to build his house at the sheltered north-eastern foot of Castle Head Hill, looking eastward along the line of his main sea bank towards Low Meathop and the rising sun. A few miles to the north-east stood the limestone massif of Whitbarrow with its white, south-facing scarp, and beyond it on the skyline the ancient rounded tops of the Howgill Fells. Northward beyond the head of the Winster Valley stood the high fells of the Lake District, hidden north-westward by the nearer slopes of Cartmel Fell.

Close about this isolated hill of Castle Head the sides and floor of the valley were verdant and well wooded. Southward the valley opened to the sea and to the wide arc of the sun for sixteen hours in summer, perhaps six at midwinter. It was, and remains, an isolated corner of great scenic beauty. Cut off from the main north–south routes of those days by a difficult journey westward across the mosses and some villainous roads, unless you came from Lancaster across the sands, and without a guide that journey was fraught with danger.

It was however a magnificent position for a house and by constructing his sea bank from the rocky hill east and west to the valley sides, John created his own spectacular seaside location. No plans of the original building have yet been found and it was substantially altered by its Victorian owner, but the place retains the calm and beauty John continued to find there until he died.

Two days before Christmas 1778 (the deadline for Abraham Darby III's completion of the Iron Bridge and a date that would draw John south again from his activities at Castle Head) he was writing an apologetic letter to his friend and business partner, James Stockdale of Cark-in-Cartmel, for not meeting him as planned before he left.[6] He wrote from The Court, sometimes interestingly just Court, now his Bersham home since the lease of Plas Grono was relinquished in 1774. Stockdale was very much his man on the spot for most of the Castle Head enterprises at this stage, acting as family friend, advisor, enabler, consultant, banker and agent by turn. John trusted him implicitly to make decisions and handle business in his long absences in the Midlands, and may even have felt under a substantial obligation to him as later events will show.

Both men had been frustrated by the difficulties of securing shipping transport for their goods in and out of the north-west ports, particularly for large iron castings that were sometimes too big to be lowered through the deck hatchways and often damaged the planking in the attempt. A letter from John's shipping agent in Chester, Hugh Jones, describes just such a nightmare loading:

> … I think in my last I mentioned to you that the Sloop was arriv'd that was to take in the Castings for Cornwall. Last Saturday I began to ship in her and have continued to do so even to this day – and not yet loaded. Yesterday I putt in the large Cylinder all safe and well, it was with the utmost difficulty got in after 5 hours attempts at it and cutting away all the deck on the larboard side, even the waterway planks, and breaking a knee.
>
> This day we have attempted to gett in the 9 foot Cylinder and have been from 10 this morning untill now past four and not yet completed in the hold – besides taking up part of the Starboard side of the Deck. There is no Room in her – the large Cylinder lays in the way of the 9 foot Cylinder – the Vessel is too small and the most unfit thing for the purpose ever yet appear'd here on such an Occasion – it's now past 5 – it's left half in and out of the hold – the Master says the smaller Cylinder must be remov'd with Screws before this can go down.[7]

In frustration John had already purchased his own ship, *The Mary*, which was by then in use on these routes. On his way south he also secured for himself and Stockdale a quarter share in the fortunes of a privateer, *The Hawke*, about to sail down the French coast round Spain and into the Mediterranean. By purchasing a substantial interest in a vessel which knew the French ports he may have been hoping to secure continuing transport for any business coming through his brother, now that England and France were at war, whatever the legal niceties might have been. There is also the possibility that he saw the arrangement simply as a speculative business venture, which could show good profits in booty. Certainly there is evidence of both British naval and private vessels preying on merchantmen returning well loaded to the ports of the European powers with whom Britain was at war, as witness the following interesting letter:

> ... A Selborne man was aboard the Porcupine sloop when she took the French India rich ship. I saw a letter from him this morning, in which he says that his share will come to £300. This will be some recompense to the poor fellow, who was kidnapped in an ale-house at Botley by a press-gang, as he was refreshing himself in a journey to this place. The young man was bred a carter, and never had any connection with sea-affairs ...[8]

The letter gives some idea of the profits John might have made for his quarter share in *The Hawke*, but this venture was a disaster. *The Hawke* was lost at sea and he carried no insurance, though James Stockdale did. He must have been convinced there was money to be made this way in time of war, however, for he took up similar options in other vessels after that:

> ... Such a train of mischief attends this war that a comparison of the evils is wanting to square them by. Not a merchant ship or Coaster sails out of Liverpool but what the owners know must sink money considerably – unless on the privateering scheme and fortunate. No prospect but in plunder which is a most unpleasing reflection ...[9]

Unless, of course, you had a share in the plunder yourself! But throughout 1779 the recurring problem for him was a consequence in part of his success as an ironmaster: the sheer volume of iron products needing transport from his several works. With water transport the only real option for longer distances it is not difficult to understand his preoccupation with coastal shipping, and subsequently, of course, with canals.

The first large order from France had in fact arrived. William had been feeding information and orders back to his brother from the beginning and had now learned of a new scheme, approved by Louis XVI, to raise water from the River Seine to be pumped along some 40 miles of iron pipes to supply Paris.[10] His meetings in Paris with the man granted an exclusive privilege by the French king to undertake this work, Jacques Constantin Perier, a well-known

French engineer, not only procured the order for pipework for William's brother in England but also brought Perier to Bersham in search of a new Watt steam engine to do the pumping. The pipework was required immediately; the engines – two were eventually ordered – could wait awhile. He added this work to the orders for cylinders and castings for Boulton and Watt and for guns for the government (now a priority in time of war). All John's works, including the furnaces at his lesser works like Wilson House, were now at full stretch making cast-iron pipes, a large proportion of 12in bore and 6ft length.

It was a frantically busy period during which a tangle of conflicting issues had to be managed. As he timed his shipments of guns for the British government from Bersham out to Chester and the Dee ports, or from New Willey down the Severn to Chepstow to coincide with the arrival of the naval escorts that provided the protection from French men-o-war round the Lizard and up the Channel to Portsmouth or London, so was he also negotiating with the French for passports to allow his vessels into the Seine with the iron pipes. There were inevitably delays in securing transport and the stacks of pipes piled and waiting on the quaysides at Chepstow and Chester led to stories, no doubt embellished in the telling, that 'iron-mad Wilkinson' was ruthless for profit – selling guns to both sides in the opposing war.[11]

Commentators have been disparaging about John's continuing to supply goods to an enemy in time of war and of shamelessly exploiting the position of having his brother close to the corridors of power in France (and, therefore, able to secure the necessary passports and access for his goods). It may be that the criticism could be more reasonably levelled at William, who had taken the Wilkinsons' Bersham gun-boring expertise to France and set up a modern cannon manufactory for the French government there to great acclaim. Admittedly, this was initially before the war had started, but it then continued under his management into the war years. Boulton and Watt were not slow to exploit these Wilkinson-French connections when they needed passports and licences for the new Watt engines, which were finding a market there. In April 1778, two months after the French had signed a supportive treaty with America in the Wars of Independence and just two months before France formally declared war against Britain, Matthew Boulton showed again the immaculate timing and judgement that had made him such a successful businessman. He secured, with William Wilkinson's help and through the good offices of his contact in France, Comte d'Heronville, a decree from the government from which the following clause is taken:

> ... The King and his Council have permitted and do so permit Messrs Boulton and Watt to manufacture, sell and distribute within the whole extent of the

Kingdom, with full agreement for a period of 15 years, exclusive of all others, the new Fire Engines of their invention assuming that the test will have been made notwithstanding, either in Paris, or in the Marshes of Dunkirk, in the presence of whichever Commissioners the Council will nominate and after which the said Engine will have been recognised as superior in efficiency and economy to the old Fire Engines …'[12]

The decree opened for them an important market in France, subject to the success of their first engines.

Early in 1779 John wrote in frustration to Stockdale in the north at his inability to get back there to further his Castle Head project: 'I am not likely to get over into the North this summer so soon as I intended. At present have no idea when I can spare so much time – not before August. With dues to you and yours remain etc.'[13] That is the whole letter. Its brevity and its clipped language provide a further measure of the time-pressures surrounding John, particularly in the context of James Stockdale's role as his agent in the purchase of some of the Cartmel Commons around Castle Head, concerning which he was in strong disagreement on valuations. He refers to a letter forwarded from Stockdale's son, acting for the Cartmel Vestry (and John) in dealings with the commissioners:

… The valuation therein transmitted appears a very extraordinary one but as I submitted that entirely to Messrs Richardson, Crossfield and yourself with the Vestry I shall abide by such opinion as they form of the valuation in question. Not doubting that the Vestry will have more candour in their determination appears to have influenced Mr Hutton and Joseph Bispham in the business. Indeed, unless they think very differently to the valuers they have appointed I shall decline enclosing the most valuable part of it next W. House rated at £16.9.2d, and as such part is certainly the best land and most convenient to me it may be considered as the best protest I can offer against the valuation made of that which is fenced off, viz.

Land enclosed – 2 acres, 3 roods, 25 perches @ 30/-	£87.3.9d
Land enclosed – 3 acres, 1 rood, 22 perches @ 4/-	£13.11.0d
Total	£100.14.9d

served in this part with respect to valuers of land for rent or purchase in similar cases which is this:- The Surveyor engages to take himself or to procure a chap at his valuation for the premises appointed to be adjudged. Now if either Hutton or his partner in this value will engage to give a rent for the same on a lease of 21 years subject to the expense of enclosing it, I will cheerfully pay 30 years' purchase as a consideration for such <u>land</u> or <u>sand</u> [John's underlining] and in this proposal it can not be suggested that I want it under its full value … However, if you with the Gentlemen already named and others that may compose a Vestry do in the meantime decide on this business, I shall certainly abide by such determination concerning that part already taken up …[14]

This letter is written in a clerk's best copperplate and provides a good example of the clerical support system John began to use as his business pressures increased, and which was to be condemned as a discourtesy by some of his later business associates. It is likely his clerk drew up the letter from a rough Wilkinson draft, perhaps jotted down in his carriage on one of the necessary regular journeys between his various works. It was a time when he would be alone to consider his problems and find solutions. It is in fact the eighteenth-century equivalent of the modern-day use of a car dictaphone, or a laptop on a commuter train. Yet in spite of the distance the use of a clerk places between the two correspondents, the letter still conveys John's position very well: irritation at an outrageous valuation; complete confidence in Stockdale's determination to do the best for him; and an absolute commitment to the purchases already made, or about to be made, by Stockdale on his behalf.

At a time when the intensity of John's business affairs in the Midland iron-making world demanded his presence there, Stockdale's role in the Castle Head project was obviously central. The two men were in regular contact by letter, but Stockdale needed him in the north more often at this important time to make decisions. In July John replied that he'd try to get there if only for a day but doubted it would be before September. In fact it was mid-November 1779 before he was able to return to Castle Head and almost a year had passed in the interim. The sea bank had been completed on time and was withstanding all weathers and tides. Land purchases and enclosures had been made in the immediate area of Castle Head and work had started on the house and the landscaping of the hill.

The importance of this visit was to assess progress made during the year and to plan the programme for the following summer. He was there for about three weeks, which included a visit to the Coniston area as the guest of Mr Knott of Waterhead – almost certainly to order slate for the large amount of roof work he was about to undertake, though also to see at first hand the Coniston copper-mining complex and assess its business potential. The visit is also significant in the context of Stockdale's report that the original roof at John's Castle Head mansion house was made of copper, though John by this time was also a friend and business associate of Thomas Williams, the Anglesey Copper King, and shipment of copper from North Wales up the Irish Sea and through Morecambe Bay directly to Castle Head itself would not have been difficult.[15]

Building a solid, four-square, three-storey structure like the original mansion house at Castle Head was a considerable undertaking. It probably required an architect and plans and it is surprising that so far such documentary evidence has not been found, the more so since James Stockdale was closely involved and many of his papers have been preserved. The original building was rendered, making it difficult to identify the source of stone, though John's close knowledge of the area would have identified good sources of building stone in at least two of the carboniferous limestone horizons outcropping within a mile or two of Castle Head. The visible coping stones and sills are likely to have come from the only sandstone outcrop in this limestone landscape, at Quarry Flat, a few miles west along the coast from Castle Head, an ancient quarry that provided in the twelfth and thirteenth centuries most of the building stone for Cartmel Priory. It was very close to James Stockdale's residence at Cark-in-Cartmel and easily accessible from Castle Head by coastal barge.

There is, however, one piece of evidence to indicate the sandstone in the building might have come from a different location. In Samuel More's journal entry for 8 September 1783, he meets again two men from Hutton Roof who had been employed as masons about the building. Hutton Roof, some 10 miles east of Castle Head, had at that time an important sandstone quarry producing freestone for the building trade.

Very satisfactory foundations for the large structure were found on the north-east side of the rocky hill at Castle Head, where the older Silurian rocks are alongside the faulted limestone. And although a large geological fault cuts through this ground not 100m from the house, modern tell-tale indicators in position for twenty years have shown no movement. John's close association with Joseph Priestley and Samuel More and other scientists of the Birmingham-based Lunar Society, many of whom were regular guests at Castle Head subsequently, suggests he would have known of the fault and taken advice on the positioning of his house on the north-facing slope where less sun exposure was compensated by better foundations and protection from prevailing south-westerly gales.

An important part of the Castle Head scheme was the landscaping of the rocky hill to form viewpoints and arbours interconnected by walks, which involved cutting the rock in places and building up the slope in others. Spiral pathways circled the hill, linked in places by rock-cut steps, all gradually ascending to a magnificent view from the top southward across Morecambe Bay to the Lancashire coast and hills. The planning and execution of this work, often in the company of James Stockdale, ran concurrently with the building of the house and continued for some years afterwards. It was a source of great pleasure and interest to John and his regular expressions of exasperation when Midlands business delayed his planned returns confirms this. There was undoubtedly, too, a deeply satisfying creative and recreational quality involved in working with nature in that spectacular location, and an awareness that the pressures of his business life needed this balance. A letter to James Watt written from Castle Head in the summer of 1780 says as much:

> … This is a very improper time of year for you to be stoved up in a room writing and drawing and by adhering to that way of life without occasional relaxation the machine for writing as well as drawing may be demolished, which will be of worse consequence than a fire burning only a part of your labours …[16]

John spent a large part of the summer and early autumn of 1780 at Castle Head, writing regularly to James Watt in Birmingham, and Matthew Boulton, gone this year to Cornwall, about the sea transport details for shipments to France and their proposed investments in the Cornish mines. His dominating preoccupation though was the building of his house and pleasure gardens:

> … I am very busy in my new box here – hope to have it habitable and convenient for a friend next summer and whish it may prove convenient for you and Mrs Watt to pay us a visit in this part – it would give great pleasure to Mrs W (who desires her compliments) …[17]

Standing tall and square above the west end of his house he built, at an early stage, a fine, detached bell and clock tower; a campanile of continental elegance and style, influenced perhaps by his brother's domicile in France or his own journeys in Europe. The tower still stands today, though it loses something of the detached strength and presence of the original by its incorporation into a Victorian extension of the house. Local folk memory has it that the tolling of the bell called from their beds in the morning the large workforce assembled by John for his building project, and the bell controlled their daily round, too. Instructions to James Watt for the clock to be directed to Mr James Roberts, upholsterer in Lancaster, were given in a letter dated 10 March 1781, which indicates that it was made in Birmingham. The broken mechanism of an ancient clock, possibly the original, can still be found in the tower today.

Something like a squatters' camp of migrant labour grew up round the hill in the summer of 1780, continuing through into the following year. It is difficult to visualise in these days of strict hygiene and building controls such a haphazard collection of shanties with pigs and poultry wandering about and crude enclosures for working draught and pack horses. James Stockdale's grandson, however, provides a glimpse into this activity:

> … Mr Wilkinson built a house on the north side of Castlehead Rock, and covered it with a novel kind of roof – one of copper: which, however, did not answer the purpose intended, and was therefore changed for a roof of lead. He covered this bare rock, in almost inaccessible places, with soil, carried up on the backs of horses in panniers, at great cost, and thus converted a barren waste into beautiful gardens and shrubberies …[18]

Samuel More, who made regular visits to stay with his friend at Castle Head throughout the building period and almost annually thereafter, has the vividness of first-hand observation in his descriptions, added to the benefit of a clear understanding of John's purpose. He describes:

> … erecting a Wall of Circular Form enclosing a Garden on the Top of the Hill. Several Fruit Trees have been planted on the outside of the Wall and are to be trained through holes left for that purpose by this means the fruit will have different Exposures on the same Trees …[19]
>
> … a considerable part of the Rock which faces the Sea to the Eastward has been covered with green Turf forming a fine Contrast to the stoney and rough places which hang over it and having mixed among the Verdure some large Stones projecting which have a very romantic and beautiful Effect … the grafts made on old and apparently decayed stocks thrive, and by way of Experiment it is proposed to engraft Liburnums and Lilacks on some of the Ash Trees and Currants on the Buck Thorn which grows wild in Abundance here In the upper Ground the Peaches and Nectarines thrive exceedingly … the Country People are continually coming to look at it and this day several People Came and when they were gone some fruit was missed …[20]

That both More and John were sensitive to the restorative spiritual power in the beauty and the silence of this place cannot be doubted. More describes 'the beauty of the Evening, the flowing of the Tide, the brightness of the Moon and the effect of a curious echo', which had an unreal haunting quality.[21] On another evening he describes the moon just rising through fleecy clouds over the eastern hills, with the western sky still red from the setting sun and the only sound the ripple of the tides, which induced in him a sense of total isolation and 'reverential Awe' from which he:

… awoke as from a Dream and recollecting myself returned to the House highly satisfied with the Sensation I had felt in my Mind during this Solitary Excursion which were truly such as I had never before experienced and which only those Magnificent Objects that every Where Surrounded me … can inspire …

It must be significant, too, that whenever John in subsequent years returned to Castle Head from his frantic business life of furnace fires and molten metal, smoke and fumes and the continual pounding clamour of his fire engines, he tried whenever possible to come over the sands from Lancaster. It may be that he touched an important spiritual quality in this crossing; an element of transition; a passing over from one life into something other; a journey into a calm and beautiful world not accomplished without difficulty and danger. As the buildings and warehouses of Lancaster faded into distance along the wide skyline behind him, the Lake District hills began to beckon ahead. More describes one such journey:

… we set out in 3 Chaises to cross the Sands to Castle Head, Miss W[ilkinson], Miss C[layton] and the Servant Maid in the first Mrs. M[ore] and myself in the second Mr. W[ilkinson] and Mrs. F[lint] in the third … Soon after we had crossed the Channel we parted from the Rest of the People who were crossing at the same Time and who were going to Flookburgh and other places in the Western Coast and turning to the Northward bent our Course toward Castle Head. Soon after we had passed the Holme Island it was necessary for us to go through another Water called the Pool which is the Tail of the River Winster, this by the Rapidity of the Freshes was cut into a deep Gully and the first Chaise which was only a few Yards before us sunk in it so much that I concluded they were all lost and was preparing to get out in order to endeavour to assist them but the Driver and Horses exerting themselves I saw them again rise out of the Water on the Sands. It was impossible in our Situation to retreat ordering therefore our Driver to Whip on we passed the Gully in the same Manner not without the imminent Hazard of our Lives the last Chaise keeping nearer the Shore escaped somewhat better, yet not without danger …[22]

It is interesting that in their responses to Castle Head and its surroundings, in their sense of identification with the place, More and John anticipated by more than twenty years a conviction conveyed in the poetry of Wordsworth and Coleridge from their own experiences in this wild and beautiful landscape, that man is richer in spirit, more whole, better able to ponder the great unanswered questions when he is open to the earth and its simple untouched beauty. The difference, of course, is that the Romantic poets embraced and lauded the world of nature unchanged. John and More improved theirs and made it fit their own view of the world in creating their version of paradise.

Not all the plans for the pleasure gardens came to fruition. The realities of that other world kept impinging. In the spring of 1781, with the planting of the hill and gardens well advanced, John realised he needed a water supply on the hilltop and wrote to Watt with details:

> … Busy as we are with engines I must have one at this place to raise water up to the top of the hill. I compute the hill (for it is not measured) at 50 yards high from low water mark. To make provision for a Jett at top to play water round the top of the hill let us call it 60 yards (rating ?) a Pump 4½ inches diameter – 4½ feet stroke. The engine will be set on a part of the hill that is perpendicular about … yards high … [Measurement is omitted from original letter suggesting either John had not calculated the height, or had not finally decided the place.]
>
> At Jack Head which will be upon the side of the hill provision to be made for discharging Salt Water occasionally – which will be wanted for a warm salt water bath – in which the hot water may be made serviceable. The use of this engine (which will only be worked occasionally in summer) is to raise water to a large cistern at top of the hill and at times to exhibit a Fountain – this will be at low water mark – when the river is quite fresh and good water – also to raise Salt Water when the tide is in – this will be very light work – and at that time a Salt Water Jet may be playd off to have a grand effect …
>
> Under these general hints what Cylinder would you recommend – and what occurs to you on the whole of this Watering Pan Scheme? This dry weather during planting on a dry hill like a Sugar Loaf which wants nothing but water at the Top to render it pleasing in many respects has put me on the plan of being independent – and nothing but a small Fire Engine can secure it … [23]

In spite of John's urgency there were delays, and the letters hint at a measure of irritation from Boulton and Watt over his long absence. Six weeks later and still at Castle Head, in a letter struggling to address their continuing transport difficulties and which manages to avoid fixing a date and time for his return, he sends Watt a reminder and further details of his engine requirement:

> … Respecting my engine to be erected here give me leave to remark that at 15 yards high I must place the house upon a precipice, must lift the water into

a Cistern at that altitude from which it must be forced to the top of the Hill
which is about 35 more or 45 yards height in all, the inclined length for the
pipes somewhere about 120 yards. At your leisure [?] you [must] not forget
this engine which I would house very compleat …'[24]

In spite of John's enthusiasm for this 'Watering Pan Scheme' the engine was
never installed and the scheme was abandoned.

The 1783 and 1784 journals of Samuel More provide this information and
a great deal more besides. They confirm the enduring friendship between
the two men and, perhaps more significantly, the respect and admiration of
More for John. That is important, for More was no fawning sycophant. 'Mr
More of the Adelphi' was a name to conjure with in late eighteenth-century
England, a man whose travels throughout the country on behalf of the Society
for Arts, Manufactures and Commerce brought him into professional and
social contact with the important scientists and businessmen of the day. It
was his responsibility to examine and assess their inventions and innovations
preparatory to recognition by, and possible grant aid from, the society.

His assessment of John in this context carries weight and provides a necessary
balance in any judgement of the man's character 200 years later, the more so in
view of the confrontations and quarrels which bedevilled John's later life. On
30 September 1783, writing in his journal and obviously sorry to be leaving
Castle Head after a month's stay, More is reflective:

> … My Friend Mr. W. who in all he undertakes endeavours at the utmost
> possible Perfection has since I was last at Castle Head continued his
> Improvements with his wonted Assiduity such as Planting Fruit Trees, Grafting
> the old Stocks, Cutting Walks, making Terraces and Walling the Garden at the
> Summit of the Hill all which seem to answer the Purposes intended very well
> and will render this Spot when completed truly delectable …'[25]

The descriptions in the Castle Head sales particulars drawn up at intervals
many years later after John's death, with due allowance for the hyperbole of
auctioneers and agents, prove that he was right.

Of the several houses John owned or leased at this time (the peak of his
prosperity and success) it was Castle Head that rejoiced his soul and gave him
the peace and relaxation that his life in the industrial world of the Midlands
could not provide. Only a few special people from that other world penetrated
this sanctuary. Mary Lee, the much-loved wife of his later life, moved from The
Lawns at Broseley to Castle Head as soon as it was habitable and was mistress
there until she died. She was a matchless hostess according to Samuel More,

who visited every year. The Stockdales lived nearby at Cark and were frequent guests. John's sister, Mary, and her husband Joseph Priestley often visited. James Watt came with his second wife, Ann, and the Flints from Shrewsbury who had looked after daughter Mary from the time of her mother's death until her father married Mary Lee seven years later.

In the early 1780s it was this same daughter Mary, now in her late 20s a well-travelled woman of poise and presence and obviously her father's idol, who also rejoiced in this new northern paradise. She stayed there for extended periods with her stepmother, Mary Lee, when her father was absent and clearly loved the place and made little corners of it her own. It may be, too, that her love of peace and beauty in this wild world of nature opened her at this important moment in her life to the love of a remarkable man, a scholar, an obscure curate from Market Drayton and the very antithesis of what her father had wished for her as a husband. It was a love that would lead her towards disaster.

8

DAUGHTER MARY

Mary was born at Rigmaden in the Lune Valley in Westmoreland and baptised on 12 April 1756 at Kirkby Lonsdale parish church.[1] Her mother, Ann, was a Maudsley of Rigmaden Hall, the first love of John Wilkinson's life. Later that summer the family moved from Westmoreland to Wrexham where Ann died on 17 November. Mary was then 7 months old, so she never really knew her mother, and the grief and misery this tragedy caused her father meant she knew little of him either during the first seven years of her life, when she was brought up by Mr and Mrs John Flint in Shrewsbury. Mr Flint was the postmaster there and a man of some standing, though what his link with the Wilkinsons was before this time is not clear. The Flints became Mary's substitute parents and obviously loved her, a love she returned until she died, but there are no recorded details of her early life in Shrewsbury.

When she was 7 years old her father married Mary Lee, a spinster aged 40 from Wroxeter, and she then went to live with them in the fine house he had recently leased at Broseley called The Lawns. It was to be her home for almost twenty years, in fact until Castle Head was built and habitable. She grew up in Broseley, where it is clear that Mary Lee also loved this daughter of her new husband who became the child she was by then unable to have herself. In spite of the tragedy of her real mother's death and the effect this had at first upon her father, daughter Mary appears to have been fortunate in the adults that influenced her early life and helped her grow to womanhood.

Mary Lee's sister, Elizabeth, had married another Broseley businessman, Edward Blakeway. They lived literally just across the road from The Lawns at Broseley Hall, so that daughter Mary's upbringing, from the time she went to live in Broseley, was one of position and privilege within the extended families of the Lees, the Blakeways and the Wilkinsons. There are letter references to family visits and outings and to a particular friendship that developed between daughter Mary and a cousin on the Lee side called Elizabeth Clayton.

Details of Mary's education are not known, nor the name of a governess or tutor if she had either. It is likely that Mrs Flint herself was responsible for her early social and domestic education and that Mary Lee took on this responsibility later, continuing through Mary's teenage years. But it is also clear from her father's pleasure in her company as she grew to womanhood, whether she accompanied him on business trips in this country and on the continent, that she had certainly been put through a formal education sufficient to understand the essentials of his affairs and that was unusual for a woman at that time. It is likely that in the absence of a son, and with no prospect of one from Mary Lee because of her age, John was schooling his daughter for a marriage within his iron-making world that would bring his business added strength and a son-in-law worthy to inherit it.

If this is so there are early signs that he was to be disappointed in his expectations. In 1775, when her father was seized with the excitement and time-consuming experiments involved in the launching of James Watt's new steam engine, Mary, at the age of 19, became engaged to a young consumptive doctor called Richard Blackley.[2] The event provides some cause for speculation in the absence of any detailed references. That Mary should agree to marry a young man so far removed from her father's ideal of a son-in-law suggests that at this point in her life she was either unaware of his expectations, or else she chose to reject them. Whereas the former is unlikely in what had become a close relationship between father and daughter, there is some circumstantial evidence to support the latter. She would see little of her father at this time and be left to make her own amusements with her extensive range of friends and family contacts. He would not, therefore, have monitored her activities as closely as at other times.

Did Mary, starved of the enjoyment of her father's attention, find a gentle young man who adored her, listened to her and made her happy? How much did her father, or indeed her stepmother, know of it? Was it a relationship of long standing that developed rapidly in this period of her father's long absences? Or was it a sudden and passionate love affair which matured quickly to a proposal of marriage and an acceptance?

Whatever the truth of this, the event did not lead at this stage, as might have been expected, to an enduring confrontation between father and daughter. The young doctor died before the marriage could be solemnised, which John might have regarded as a blessed release. His correspondence for 1775, the year daughter Mary became 19, is sparse but contains no reference to a Dr Blackley, which is surprising for an event of such significance in the family. However, on 14 July 1776 Samuel More, visiting Broseley as John's guest 'found the family much distressed, a Young Gentleman [Dr Blakeway] being extremely ill and his death expected every Day'.[3] John records the death in a business letter to Matthew Boulton, dated 9 August 1776, apologising for the consequent delay to their meeting: 'We are in great distress here on the loss of Dr Blakeway who died yesterday ...'[4]

It is tempting to assume that this is the Dr Blackley engaged to Mary and that the difficulties of handwriting had produced an alternative transcription. The young Dr Blakeway was of course Mary's cousin on her stepmother's side. The Blakeways lived at Broseley Hall, very close to the Wilkinsons. If this were the man to whom Mary had become engaged the two would have known each other from childhood. The consumption might have developed late, and rapidly; it is unlikely there would have been any parental objection to such a marriage. This could explain why there was no enduring confrontation between father and daughter about what, in her father's view, would nonetheless be an unprofitable match ... but all this is speculative.

During the next few years there is good evidence of increasing closeness between father and daughter, and particularly at the time of his land purchases in the Castle Head area in the late 1770s. In the summer and autumn of 1778 they are all there together – father, daughter and stepmother – involved in the excitement of John's plans for the building of his northern sanctuary. They are there continuously for four months staying at Wilson House, the Wilkinson family home when John was a young man and which had obviously remained in the family following the earlier move to the Midlands. It subsequently became the home farm of John's Castle Head estate. Daughter Mary loved the area, which would be important to her father, and Samuel More's journal tells of her labours to create a garden and grotto of her own in the rocky outcrops close to Wilson House.[5]

Throughout this long stay in the north John remained in regular contact by letter with Boulton and Watt and his business affairs elsewhere in the Midlands, but the need of his presence there eventually became essential. Leaving Castle Head affairs under the supervision of his trusted friend, James Stockdale, he travelled south just before the Christmas of 1778 with his wife and daughter. Mary left them in Warrington to visit friends in Liverpool. He and Mrs Wilkinson journeyed on to his house at The Court, close to the Bersham works, before returning to Headquarters at Broseley early in the New Year where Mary eventually rejoined them. In the interim her father had bought a ship in Liverpool, frustrated by difficulties and delays in the shipping of his large cylinders and castings up and down the west coast and through the Channel. It must be significant that he called her *The Mary*, and perhaps an omen that she was to be involved in uncertain voyages to the French coast in the following year.

During the spring and summer of 1779, daughter Mary's time seemed to be divided between Broseley and Shrewsbury, where she obviously had another close circle of friends. For the first time there were hints of an uncertainty and restlessness in her behaviour, noted by her father in his letters: 'My daughter is now at Shrewsbury – has some thought of going to France if proper company offers – but there are so many difficulties that it's much if she can get away this year ...'[6]

John was prevented by business pressures from returning to Castle Head for a long stay that summer, anxious though he was about the work afoot there, but he managed to escape for a three-week visit with Mrs Wilkinson in November and December. Mary did not accompany them. References to her in the letters indicate she was away from home visiting friends.

They were back at Broseley by Christmas and during the spring and summer of 1780 John's frustrations and preoccupations became near desperate in attempts by himself, his brother William (now resident in France) and Boulton and Watt to obtain documentation from both French and English governments, at the time of hostilities between them, for shipping engines and engine parts to a French port. Hurrying from one business centre to another and needing to maintain contact with Boulton and Watt, he outlined his intended movements carefully, which tells us that on Monday 22 April 1780, on a journey between Bersham, Snedshill and Broseley, he will 'return home by way of Drayton where my daughter is upon a Visit at Mr Feltons'.[7]

Henry Felton was an attorney-at-law in Market Drayton. It is the first time that Drayton, or Market Drayton, has been mentioned as a focus of Mary's friendships and in view of what is to come it is important. It seems that her father simply picked her up from there to take her back home via his works at Snedshill and nothing further was mentioned of that visit.

That summer the Wilkinsons returned to Castle Head for an extended stay. John was still frustrated in the business of passports for goods to France, and also in extending what they all believed would be very profitable engine business into the well-established mining areas of Cornwall. It was not a good time for him to be distant from his business world, which is a measure of the importance and enjoyment attached to his project at Castle Head, even without his daughter, who did not join them there on this visit until the end of October. She had been almost three months without them, either at home in Broseley or with friends. The family returned to Bersham in time for an important meeting with James Watt at the end of November, but not for long.

By early February 1781 they were at Castle Head. From the absence of any reference to his daughter in the compliments he included to the Watts family at the end of subsequent letters, it seems likely that daughter Mary did not accompany them. This was the year when John had hoped to have his new residence finished and habitable, and he had already invited the Watts to spend some time with them there in the summer. It is clear he planned to be there from early spring and to stay there until the house was ready. Soon after his arrival, however, there was an unavoidable diversion to Kirkby Lonsdale. Writing to apologise to Boulton and Watt over delay in answering their business letters, he explains: 'Gentlemen, The sudden Death of an old Lady, Grandmother to my Daughter, about 15 miles from hence … has involv'd Me in additional Business and occasion'd my being absent from hence 3 Days …'[8]

The old lady in question was Margaret Maudsley, the surviving widow of Thomas Maudsley and by then the mistress of Rigmaden Hall. In his next letter a few days later he explained further: 'She has left us without a Will and being an undivided Manor and Estate in which my daughter has one half, the active part in adjusting her affairs falls to my part …'[9] John, in spite of the other pressures on him at this time, did what was necessary to secure this inheritance for his daughter. On 4 April 1781, at a formal meeting held at Rigmaden of the Court Baron and General Court of Dimissions, the necessary general fine due upon the death of Margaret Maudsley to secure the succession of her inheritance to her granddaughter, Mary Wilkinson, and to her only surviving daughter, Mrs Margaret Robinson (a widow and Mary's aunt), were levied on fourteen tenants of the estate and the money paid over by the court steward to John Wilkinson.[10] There is no direct evidence that Mary was present on this occasion, nor indeed that she was at Castle Head at all throughout this year.

Urgent business took John back to Bersham, Broseley and Birmingham during the late summer, but he hurried back to Castle Head to supervise the finishing work at the end of September and remained there for another six weeks. From the letter references it seems that Mrs Wilkinson stayed behind at Castle Head in charge of affairs during this brief excursion south.

The next mention of daughter Mary is in a letter to James Stockdale just after Christmas.[11] It was written from London where John had arrived after an excursion into Cornwall: 'My daughter is very well and sends dues. We think to leave town on Friday if weather permits. I hope we shall in a few days be on the other side. My Headquarters will be in Brussels …'

The decision for father and daughter to travel together to the continent at this time of unrest in France will have been carefully considered. Mary's restlessness and her expressed wish to travel in France will have been part of it. It would also be typical of her father, in his present difficulties with passports for engines and pipework, to want to assess the problems on the other side for himself. The best accommodation for them both would be to travel together, with the added insurance of the company and the fluent French of brother William, who met them in Ostend to accompany them as interpreter. John expected to be there until March, though he was not impressed with what he found: 'Nobility Priests and Beggars constitute the greatest part of the people here …' Though he also noted that: 'Here appears to be a very large field open for Fire Engines at some future date … They are at present in a most miserable situation in these engines … not equal to the <u>worst</u> [John's underlining] in England 20 years ago …'

John also mentions in a further letter that his daughter 'has met with a very decent young woman of this place for a servant yet as she only speaks French and Flemish Mary must set about learning French in good earnest …'[12] Uncle William, much nearer to her in age than her father as a bachelor aged 38 and resident in France, would almost certainly have been her tutor. Mary was 26.

There was an absence of Mrs Wilkinson's compliments at the end of letters written from Broseley in the spring and summer of 1782, which suggests she had remained at Castle Head to supervise the fitting out of the house. John was involved with the itineraries of visiting French and Swedish ironmasters at this time, and preoccupied with the business of bringing his new Bradley works into production and the teething troubles in setting up a Hammer Forge there powered by a new Watt engine. By early October, however, he was back at Castle Head; flying south again to attend to business before returning to Castle Head for Christmas.

There was no mention of Mary in his letters until 5 March. Writing from Broseley to James Stockdale at Cark there was a revealing postscript: 'Last letter from my daughter and my brother since I came here is from Genoa, February 3rd – all well.' There had obviously been earlier letters, but what is not clear from the evidence is for how long had they been travelling on the continent together? Did Mary remain behind with William when her father returned to England from Brussels in the spring of the previous year? Taking advantage of her uncle's protection for the excursion in France she had said, two years previously, that she wanted to make? And did her father approve this arrangement, as a feminine equivalent of the Grand Tour that would increase the poise and presence of his daughter and enable her to learn French? Had the two been together for the whole year?

It is important to remember here that William, aged 39, was sixteen years John's junior and perhaps his stepbrother rather than a full brother. Since he went to France in 1776 he had built, from virtually nothing, a modern cannon forge at Indret for the French Government and was respected in France and very well paid. As a young bachelor he would have acquired French poise and polish with his increasing fluency in the language, and he would be a very attractive travelling companion for the 27-year-old Mary. It is tempting to speculate that their relationship might have developed through a long period of close contact into something deeper. The bitter quarrel between the brothers, which developed some three years later when William returned from France, is evidence that could support this. But the evidence of the letters of the time is that John and William were still on friendly terms in that period following Mary's return from France.

John, however, was clearly anxious about his daughter. In July 1783 he extended a business visit to London, a place he did not like, 'not knowing how soon I may be summoned across the water to meet my daughter ...'[13] There was an arrangement in place for William to bring Mary to Calais and for her father to meet them there to bring her home.[14] He did not have long to wait and, much relieved, was back in London with his daughter some ten days later: 'Mr More attended me to Calais and we had a very agreeable time there four days, a pleasant passage there and back and all safe back the eighth day from our leaving London ...'[15] This same letter also contains mention of business matters discussed with his brother and suggests that the reunion in Calais had been a happy occasion and the brothers had parted friends.

On 7 August Samuel More and his wife joined John and his daughter on the journey home from London on the post coach. There was a celebratory reunion in Birmingham with the Priestley branch of the family, together with the Watts and other of John's old friends from the Lunar Society. There was a party feeling about Mary's return, which continued the next day on arrival at Broseley. There, along with her Uncle Blakeway and his new young wife, Mary's close friend and cousin Elizabeth Clayton waited to welcome her. A couple of days later the Flints came over from Shrewsbury; there were dinners with the Reynolds and plans were laid for them all to join Richard Reynolds' annual picnic on the Wrekin the following week.

Arriving in Broseley on 15 August to join in these celebrations came a young man, a friend of Richard Reynolds who for the next two years became central to the Wilkinson story. Samuel More described his arrival: 'This Day Mr Holbrooke a Clergy Man from Draiton came to Broseley and with him our Afternoon was spent agreeably. He is an intelligent Sensible Man and has withdrawn himself from the Church of England on Account of his not approving the Doctrine of the Trinity.'[16]

Theophilus Holbrooke was much younger than the Quaker Ironmaster and it was at Reynolds' invitation that he had come to Broseley for the annual Wrekin picnic. The friendship was clearly of special significance for Reynolds. In a handwritten, personal list of business deals, purchases and family births, marriages and deaths Reynolds recorded: '1783 Acquaintance with Theo Houlbrooke commenced.'[17]

The two men were both thinkers and intellectuals; both interested in religion albeit from different bases; and lived quite close to one another. Was it the Quaker Reynolds who had led the young Holbrooke to question the Church of England doctrine of the Trinity, a process with which John himself perhaps would have had some sympathy? There is no evidence, however, that John and Holbrooke had met before the day of the Wrekin picnic.

There was an excited, happy atmosphere about this occasion, which More captured in detail in his journal as the various participants arrived at the top on foot or on horseback from all points of the compass. John and his daughter were in the Reynolds' party which included Holbrooke. 'We sat down to the cold Collation which was spread on the green Turf and regaled ourselves heartily with it …'[18] A tour of the summit followed and Richard Reynolds and Samuel More demonstrated an easy way to save the 4-mile walk back to the waiting carriages by sliding down the steep mossy slopes on their backsides. It must have been fun.

Holbrooke made a special appearance at the New Willey furnace on the evening of 20 August to take leave of John and More before he set off back

Richard Reynolds, ironmaster of great influence at Coalbrookdale and Ketley.

to Market Drayton early the following morning. A couple of days later John and More left on horseback for Bersham en route to Castle Head, then on 1 September they were joined at John's house, The Court, by the ladies who were to accompany them: Mary Wilkinson and her friend Elizabeth Clayton, and the two older ladies Mrs More and Mrs Flint. Over the next three days the party travelled on to Castle Head in two chaises and were given a great welcome there by Mrs Wilkinson on arrival.

It was during this visit, on their way to Cark for breakfast one morning, that John discussed with More, and probably with James Stockdale too, the question of a relationship between Mary and Holbrooke. The reference in More's journal is teasingly spare: 'It was in our Way hither that Mr W. first mentioned to me any Thing relating to Miss & Mr. H.'[19] The tone of More's brief remark suggests there is something in the wind, and it is the first clear evidence of a concern by John over the behaviour of his daughter.

The party stayed at Castle Head until the end of the month and Mary and Miss Clayton were not with the others when they left. The two young women had either left earlier or else they remained behind with Mrs Wilkinson. Either alternative suggests a rift in relations between Mary and her father. His was a very direct, even aggressive, personality and it is likely there were serious tensions between them during this visit, perhaps even open conflict. From what is known so far of Mrs Wilkinson she would have tried to mediate. Mary would have been grateful to have the support of her close friend, Elizabeth Clayton, Mrs Wilkinson's niece; John glad of the opportunity to discuss his problems with trusted friends of long standing. He knew how tensions like this were likely to affect him: 'Peace is a most desireable thing and the more so to one of my constitution who cannot be angry by halves. Resentment with me becomes a matter of business and stimulates to action beyond any profit whatever ...'[20]

Three business letters written during this period by John to James Watt in Birmingham are revealing.[21] They are different from his former, friendly letters to Watt and have a restless, irritated, even carping tone about quite small irregularities and changes in his works, with hints of injustice and a determination not to be subject to the persuasions of others. The specific compliments of Mrs Wilkinson and his daughter to the Watt family, included in all previous letters when his family were with him, are missing.

The same tone continues in subsequent letters to Watt, and at the end of what is clearly an unsettling time for him, when for some reason he must remain at home at Broseley until 6 November, and following a further period of business frustrations and conflicts, he wrote: 'We come at part of the truth, but the whole is wanting to form any judgement by ...'[22]

John was back at Castle Head for about six weeks over Christmas, accompanied by his brother William. Mrs Wilkinson had obviously remained there, and it becomes clear that she now regarded this new residence as her home. There was no mention of Mary until the following August when she returned to Castle Head for a summer visit. A brief reference in a letter of 1 August suggested she had followed her father to Castle Head and did not travel with him. Nor did she accompany him south on 5 September when he met Samuel More and Thomas Williams, the Copper King from Anglesey, in Wigan for an excursion into Wales.

It is probable that Mary took advantage of her summer excursion to the north to visit Kirkby Lonsdale and Rigmaden, and perhaps to see her aunt, preparatory to the disposal of the Rigmaden estates which had now been posted for 28 October 1784 and in which property each held a half share.[23] Her father clearly planned to attend this meeting, which was perhaps the reason he cut short his Welsh tour and took ship from Anglesey for Liverpool, shortening the ongoing journey to Castle Head by chaise. He, too, would need to familiarise himself with the sale documents and arrangements. Samuel More travelled with him for this brief ten-day visit to Castle Head but, unusually, found his own entertainment: visiting Kendal and Barrow without John who clearly had more urgent business to attend. The two men left by chaise on 29 September. Again Mrs Wilkinson remained behind. Mary travelled with them for the two days to Wrexham but 'was in haste to get to Shrewsbury' and obtained a lift onwards in the carriage hired by a lady known to More.[24] After two days' close confinement in a chaise with her father it seemed she was anxious to part company with him.

More gave no details in his journal of what passed between father and daughter, but he would have been party to it all. Again there are only teasingly brief references. Following her flight from Bersham Mary had taken refuge with the Flints in Shrewsbury, where the two men caught up with her four days later and obviously persuaded her to travel on to Broseley with them in the post chariot. She did so but insisted they call on her friend Elizabeth Clayton in Wroxeter on the way. After dinner together at The Lawns that evening Mary again made an escape, this time across the river to The Dale and in all likelihood to the Reynolds' house where Holbrooke would be waiting.

Over the next four days John immersed himself in his New Willey furnace affairs and did not accompany More on what were largely social visits in the area, though they were both staying at The Lawns where Edward Blakeway joined them for dinner in the evenings. The talk of the three men would have been interesting. It is possible to catch something of John's frustration with the Quaker intellectual world in which his daughter had become involved in

William Reynolds, son of Richard Reynolds, developer of industry at Coalport.

an echo from More on a visit to the small cottage where he 'met Miss Hanah Reynolds who here indulges herself too much in Contemplation when her Accomplishments and beautiful Person should lead her to Shine an Example to the other Young Women of her Acquaintance'.[25]

This of course was Richard Reynolds' daughter, but could More have had Mary Wilkinson in mind, too? His last mention of her comes on the day he left Broseley at the end of this visit, John accompanying him in the post chaise as far as Birmingham. It was a last opportunity for some time for the two friends to talk confidentially together: 'Passing over the Wooden Bridge the Conversation turning on the Conduct of Miss W ...'[26] The Wooden Bridge was the bridge over the Severn, downstream of the new Iron Bridge, which then provided the shortest route from Broseley to the Reynolds' property on the other side.

If Mary Wilkinson was in turmoil because of her feelings for Theophilus Holbrooke and the consequent conflict with her father, Holbrooke was equally disturbed. Earlier that year his written correspondence with Richard Reynolds, who was away from The Dale at the time on his summer excursions, explored ideas concerning Christian understanding, happiness, friendship and love. Unfortunately only Richard Reynolds' side of this exchange has been preserved, but he was responding to ideas raised by Holbrooke and extends and deepens the intellectual discussion.[27] Reynolds' letters clearly indicate first that Holbrooke was in a state of rapturous happiness but was troubled by an accompanying shadow. There is also evidence that he had confided in Reynolds about his love for Mary Wilkinson, which provided the context for these replies. They show a closeness of spirit between the two men and a massive respect for Holbrooke from Reynolds. They also reflect a wide reading background and show Reynolds to be a master of the English language, capable of expressing complex ideas fluently in excellent prose.

If the shadow to Holbrooke's happiness was John's response to Holbrooke's love for Mary then Reynolds would have some difficulty with this. He had been a friend of John since his widower days back in the early 1760s. These two extremely capable ironmasters had a respect for each other born of experience and long association, as every known contact between them in the interim indicates. Yet Reynolds was about to support Holbrooke against John. It was a decision that would have involved him in mature reflection, even prayer, and is a measure of his regard for this new young friend.

The projected meeting at Kirkby Lonsdale on 28 October 1784, in which the succession to the Rigmaden estate was passed equally to Mary Wilkinson and her aunt, was obviously completed satisfactorily since the following summer Mary's title to a half share in this property features in a new indenture.[28] It is a very informative document, dated 23 July 1785, and shows that by then Mary and Theophilus Holbrooke were engaged to be married. Mary's share in the estate was clearly intended as her marriage dowry and by this indenture was passed in trust to Richard Reynolds and William Tayleur of Buntingsdale in Shropshire, Holbrooke's long-standing friend, until the marriage was solemnised. There is no reference to Mary's father in the document.

It becomes clear that John's opposition to this love match had reached the point where he refused to support Mary financially in the marriage, and that Mary, equally obdurate, had found a way to use her Rigmaden inheritance, which came to her quite independent of her father, for this purpose. She also had the support of influential friends to help her to implement this: Richard Reynolds of course, but also Henry Felton, attorney-at-law of Market Drayton with whose family Mary was on visiting terms, and two further attorneys from that place. Henry Felton and a mercer in Market Drayton called Thomas Grant witnessed William Tayleur's signature as trustee in the indenture, and two ladies of the Reynolds family, his daughter Hannah Maria and his cousin Mary Ann, witnessed that of Richard Reynolds. It is likely that because all the signatories came from the Market Drayton area the document was drawn up and signed there and was not dealt with at Kirkby Lonsdale. Whether John knew of it in advance is not known.

Theophilus Holbrooke and Mary Wilkinson were married without banns by special licence on 8 October 1785 in the diocese, if not in the parish, where at one time he had been curate.[29] It seems to have been a lonely affair. William Tayleur was there as witness for his friend. The minister's wife witnessed Mary Wilkinson's signature – which suggests she had no one standing beside her. Perhaps the whole occasion was kept quiet to make it difficult for Mary's father to arrange a protest.

Three miles from Market Drayton lies the pretty hamlet of Moreton Say, with its unusual church and delightful vicarage and garden just across the road. It is the place where, in 1781, Theophilus Holbrooke, an Oxford scholar but a Shropshire lad himself, began his Ministry. In spite of his subsequent lapse from grace over his difficulties with the Doctrine of the Trinity, and since the tragic sequel to this marriage is focussed here, it is tempting to assume that following the marriage he chose to return to this idyllic place with his young bride.

The transfer of the dowry from the trustees to Holbrooke took place the following February, by which time Mary was heavily pregnant.[30] They had by then decided to sell Mary's half of the Rigmaden inheritance rather than to keep it and use the substantial rents as income. Mary's Aunt Margaret had made the same decision, though there is no record of discussions between them on this matter. A buyer was found called John Satterthwaite from Lancaster. He wished to have the whole estate and the transaction to purchase the two halves together was completed on 10 February 1786. Mary's share in the proceeds amounted to £3,900, which would have solved any immediate financial difficulties they might have had following the withdrawal of her father's support.

For long periods in the summer and autumn of 1785, and through the winter and into the spring of 1786 John remained at Castle Head with only short trips south around 'Quarter Days' on business matters. The Wilkinsons were entertaining family and friends in their new residence during much of this time. Ann Watt made a brave journey south from Glasgow with her children by post chaise to see for the first time in months her husband, who travelled up from Birmingham to meet her at Castle Head. John informs his friend of their safe arrival: '5-o-clock – Mrs Watt and Co are just arrived – all well and will not be permitted to quit this Castle until you come and set them free …'[31]

The Priestley family, also from Birmingham, were there too, and there must have been great reunions and rejoicings during that winter of 1785/86. Mary is never mentioned in the letters and it may be that John's long residences at his northern sanctuary throughout this time were an escape.

Mary Holbrooke, née Wilkinson, gave birth to a baby girl the following May. The child died immediately after birth and was buried at Moreton Say on 25 May 1786.[32] Imagine the contrast of this bleak misery with the love and hope the young couple had brought to this beautiful place, fragrant then with flowers and May blossom. They named the dead child Mary. Worse was to come. There must have been complications at the birth for in less than a month Mary the mother was also dead, buried with her dead child at Moreton Say on 18 June 1786.[33] A grieving Theo placed a stark memorial tablet recording the details simply in Latin on the external, sunless north wall of the church. It remains there today, forever in shadow, with space below the names of his wife and daughter for his own to be carved one day.

There is an interesting discrepancy in the evidence relating to the death and burial of Mary and her baby. The Market Drayton church registers record that the infant daughter died immediately after birth and was buried at Moreton Say on 25 May 1786, and that Mary was also buried there on 18 June 1786. Theo's memorial tablet, in classical Latin with standard abbreviations, has a different story. It says that a funeral service took place in which Mary and her dead baby were interred in the same tomb at the same time.[34]

No one knows if Mary and her father were ever reconciled in the month that she lay dying, or indeed if John and Holbrooke, who lived on for more than thirty years, ever spoke to each other again. There are clear similarities in the fortunes of both men in their first marriages and the present tragedy must have woken in John memories of his early life. Both men loved their wives deeply and after a brief period of happiness both lost them soon after childbirth. Both met family resistance to their marriages. John's marriage lasted just long enough for him to overcome this and be reconciled with his first wife's family. The Holbrookes met harder resistance and their marriage was brief. The tragic irony for John was that his child had survived the first tragedy to become the principal in the second. Did he ever ponder these things? Was he sorry to lose, in the prime of her womanhood, this strong-willed daughter, the one-time light of his life? Or had he, as with other uncomfortable things around him, cut himself free of her to concentrate on what he could influence and change?

Whatever the truth of this, his long friendship with Richard Reynolds continued and there is no evidence that John held any grudge against Reynolds for the part he had played in assisting the Holbrooke marriage. Did he forgive Reynolds, or was forgiveness not an issue? Perhaps he understood that there is a hierarchy of friendship and Reynolds' loyalty to Holbrooke was higher than any loyalty to him. John was always a pragmatist even in matters close to his heart, and it is likely that after this tragedy he paused and reflected for a while, was sad yes, but then set a new course and started his life again.

He was now 58 years old and this time it was not so easy. He was still without an heir and had lost the means of begetting one of his own blood. As his business world expanded the pressures on him increased and the incidence of frustrations and quarrels within that world became more frequent. The sale of engines to Cornwall and the devious manipulations of the Cornishmen was a particular source of friction, and the delay in supplying Watt engines, particularly to his friends when he as the founder was supplying all the parts, clearly infuriated him. Boulton and Watt continued to refuse him a partnership in the new engine enterprise, and though he still regarded Watt with affection there is evidence of a new friction with Boulton. A letter written shortly after Mary's death, enclosing an overdue account from Cornwall, shows a different John:

> Messrs Boulton & Watt, Gentlemen,
>
> Be so good as to Attend to the above Accounts and Dates with the Balance due thereon. Your Mr Boulton can recollect what I said upon this scrawling business when he and I were together at Marizion. If he does not I will here repeat it – that after such Treatment I could never Trust my property a second time with the Adventurers in Hallamanin or Wheel Union ...

And on a specified delay in providing engine parts:

> I need not mention to you the reason why these last were not ready sooner. They have wrought at Bersham night as well as day, Sundays included. They continue to do so and your last orders will be expedited as fast as is in our power – and as it is so much easier for Mr B to execute than to order I shall beg the favour of a lecture from him in the Foundry line the next time I see him ...[35]

There is anger and unveiled sarcasm towards Matthew Boulton here, and a sense of struggle and injustice as he defended his workmen and procedures against unreasonable complaint. It marked a change in his relations with Boulton and Watt, and the beginning of a determination to depend no longer on others but to bring all aspects of his iron-making manufacture and sales under his own control. It was a hinge point in his life at which he needed to make a statement about himself and his work as deteriorating personal and business relationships threatened to confound him. That statement came in the form of an iron boat.[36]

As early as 1777, at which time John's life was almost consumed by the process of refining the first few of James Watt's new fire engines, he had noted the successful use of a small iron pleasure boat on the River Foss in York, enthusiastically reported in the local newspapers. That information appears to have surfaced again in 1785 or 1786 when Stockdale reported, using letters written from John to his grandfather (which he then held but which have since been lost), that experiments were going forward at the Bradley ironworks to build an iron boat. That timing is concurrent with the quarrels and traumas in John's personal life described above.

To build his iron boat at the Bradley ironworks with immediate access to the Birmingham Canal made sense if he was building a canal boat, a narrow boat something under 7ft wide, but not so much sense if he was building a river barge with twice the beam for use on the River Severn. Since it was a narrow boat that he successfully launched the following year on 9 July 1787, but on the river at Willey Wharf rather than on the canal, it must be that for some reason his Bradley ironworks were not as suitable for this experimental building project as his New Willey works, which had iron rail access to the Severn at Willey Wharf. This first iron boat was instead transferred from the river to the canal at a later date.

By the end of the following year, however, three more iron boats had been launched; one of them certainly a river barge which was reported to be still on the Severn in 1803. That sighting is evidence that John's iron barge at least was able to withstand the heavy stresses of grounding in the shallows and of being beached, which reduced the useful life of the other iron boats by then appearing on the river. It is clear that the canal boats, which suffered much less damage in this respect, had a longer working life.

The pride with which John reported the launching of these boats to his friend James Stockdale of Cark near Castle Head is evident:

Broseley, 14th July, 1787.
Dear Sir,
    Yesterday week my Iron Boat was launched. It answers all my expectations, and has convinced the unbelievers, who were 999 in 1,000. It will be a nine days' wonder, and then be like Columbus's egg ...[37]

And later, apparently in response to a request from his friend for information:

Bradley Ironworks, 20th October, 1787.
... There have been two iron vessels launched in my service since 1st September, one is a canal boat for this navigation – the other a barge of 40

tons for the river Severn. The last was floated on Monday, and is I expect now at Stourport with a lading of bar iron. My clerk at Broseley advises me that she swims remarkably light, and exceeds even my own expectations ...[38]

The evidence from these letters is that these were John's first iron boats. We must, therefore, seriously question claims made eighty years later by Stockdale's grandson of an iron boat on the river Winster at Wilson House near Castle Head in the 1750s. It is also significant that modern attempts to locate the earlier boat using sophisticated search equipment, co-ordinated by the Windermere Nautical Trust, have produced nothing.

The impact created by these iron boats put John again in the forefront of new iron-making technology in the hub of the growing industrial world, but more than that it demonstrates again an important aspect of his character. When his personal life met with tragedy he buried himself in his iron-making business where he was certain of himself, his experience enormous, his touch sure and his opportunity to experiment personally fulfilling and reassuring. At this point he was probably at the apex of his power and influence. The succeeding years found him in increasing disagreement with his contemporaries, with further quarrels and the beginnings of decline.

9

BROTHER WILLIAM AND FRANCE

The question has been posed by Challoner, in the absence of conclusive evidence to the contrary, as to whether William Wilkinson was John's full blood brother of the same parents, or indeed a stepbrother with a different mother.[1] The apparent twelve-year gap between Henry and the next sibling, Mary, and the possibility that Henry was damaged in some way thus discouraging further child-bearing, is the cited evidence for this. It suggests that Isaac's first wife had died sometime after John and Henry were born and that a long barren period followed before he married another woman and emerged into the sunlight again.

Some recent research by Janet Butler, however, has constructed a family tree that indicates that no such twelve-year gap occurred. Rather, in that period two further children were born to Isaac and his wife.[2] In the settlement details following Isaac's death, his wife is listed as Mary, formerly Mary Johnston, and a daughter Margaret is included. Also, 'my late sister, Margaret' is subsequently mentioned in the will of his son, William.[3]

Isaac's Backbarrow years, after the birth of John and before Mary, were busy years in terms of his work as an emerging ironmaster. It was a period of great inventiveness and energy when his new box iron appeared and when he erected the first iron bellows at the Backbarrow Company's furnace and forge to the acclaim of his employers.[4] It may be, of course, that the pattern of behaviour which emerged later in his son John at times of great personal stress in his life – when he poured all his creative energy into his iron-making world and stood back from close personal contact with family and friends – was a pattern transferred from father to son and here exhibited by Isaac. However, to date there is no more than circumstantial evidence for the death of a first wife, and no record of a marriage to a second. That which might have been his first and only marriage, to Mary Johnston, endured into old age.[5]

It yet remains clear that John and William, who in their mature years were each to make so great an impact on the eighteenth-century iron-making world, grew up apart as a consequence of the sixteen-year age difference between them; sharing little or nothing of their childhood and teenage lives. This may be more significant in their subsequent relations as men than whether they were full brothers or stepbrothers.

William's childhood years were spent at Wilson House in Cumbria when Isaac moved there from Backbarrow in 1748. Here he grew up with sisters much closer to him in age than John, who by then had almost certainly left the family home anyway. When Isaac moved to Bersham with the rest of the family in 1753, John remained behind for a further three years at Kirkby Lonsdale during which time he married Ann Maudsley and established his own iron-merchant business there. As William entered his teenage years at Bersham in a family dominated by girls, it is likely that he enjoyed more of Isaac's favours and attention than either his sisters or his absent elder brother. Isaac subsequently sent him to school at the dissenting academy of the young Reverend Doctor Joseph Priestley who later married his eldest daughter Mary and who provided William, as Isaac clearly intended, with the best education available to a dissenter outside of the Church-controlled schools and universities. As a stripling of 16 or 17, William then returned to Bersham to learn iron-making under his father and his elder brother, John, who by that time had followed Isaac into Denbighshire from the north.

The relationship between the brothers at the time, with John twice William's age, would be that of man to youth, teacher to taught, with John by then a widower and a man of some worldly experience, but perhaps a touch jealous of the favours and attention bestowed on William by their father. There is, however, no immediate evidence of a rift between the brothers, rather the reverse. Though John was now beginning to forge a way for himself in the iron-making world independent of his father he clearly relied increasingly on William to manage affairs at Bersham to their mutual advantage, and when Isaac left Bersham a few years later the business was re-established without him under the joint control of the two brothers as the New Bersham Company.[6] William was not immediately accepted as a partner in the New Bersham Company and had to wait until 1774 for this preferment, when he was granted only a ⅛ share by his brother.[7] It is from this later date that their differences began. That there was a quarrel in the earlier separation of father and sons is certain, though it appears to have been between Isaac and the sons, certainly between Isaac and John, and not at this time between the brothers.

The beginnings of this quarrel are found in a legal battle[8] between Isaac and one of his partners, his nephew William Johnston – the son of his wife's younger brother, in the company first set up by Isaac and then called the Bersham Ironworks, over shares purchased in 1761 from William Johnston by Isaac but never satisfactorily transferred to Isaac's name.[9] The dispute became so bitter and protracted it discouraged two proposed new partners from joining the company, thereby depriving it of further capital that was clearly needed at the time.

Sometime after joining his father at Bersham in 1756 John became 'a principal manager and acting person in the affairs of the said company', though not a shareholder until 1763.[10] In that year, seven years after the death of his first wife, he married for a second time. His new wife, Mary Lee, was already a shareholder with Isaac in the old Bersham Ironworks and since her shares transferred to her husband on her marriage John thus acquired her one-third shareholding in the company, equal to that of his father. It may be that this was the trigger to the subsequent conflict over shares between Isaac and William Johnston in which Isaac, through the acquisition of these further shares, fought to retain control as principal shareholder of the company he had founded. John will have pointed out that these shares had never been properly transferred to Isaac so that in the event of any disagreement between them his father's vote carried no more weight than his own. It is another early example of the shrewd business manoeuvring that came to be associated with John.

William Johnston held the remaining one-third share in the company equally with a fourth partner called Samuel Green. It is clear from the incomplete chancery record that William Johnston believed it was Isaac's conduct and temper that led to the breakdown of the old Bersham Company, but it is not clear how far William Johnston's sympathies lay with John nor how far the dispute might have been provoked and engineered by John in an attempt to gain control himself. As cousins and near contemporaries, John and William Johnston would have something in common, and as John's wealth and influence grew in later years William Johnston's name certainly recurs as a trusted friend and employee within his business empire. If Isaac believed he had been betrayed by them then the basis was laid for future uneasy relations between father and son, and for the litigation which eventually followed.

How far Isaac had encouraged the marriage between his son and his business partner, Mary Lee, through those seven years of John's grieving for his first wife also has to be speculative. But it seems clear that Isaac had known Mary Lee from his earliest years in Bersham, that she was his business partner from the beginning of the old Bersham Ironworks and that he knew her long before his son did and might have introduced them, however much the subsequent marriage worked against his own interests.

Through the late 1760s and early '70s, William Wilkinson's name does not feature in these family quarrels and it seems he immersed himself in the business of learning the secrets of iron-making and the iron trade, with some success.[11] These are his salad days, his years between the age of 20 and 30, but no woman emerges during this time as a significant influence or companion and little is known about his personal life. It may be that, in the single-minded way of a Wilkinson, he had dedicated himself to becoming as competent an ironmaster as his brother and his father and had put aside the indulgences until later.

His immediate and enthusiastic reception at the end of this period of Marchant de la Houliere's proposals for him to go to France and build a new iron foundry for the French government suggests that he recognised his time had come.[12] He would be aware of the wider social opportunities the invitation opened to him, but perhaps more important was the chance to do something in his own right away from the dominant influences of his elder brother.

In a first contract with the French government, which carried the hallmark of a seasoned negotiator and suggests that brother John had been involved, William committed himself for no more than a two-year period initially.[13] The contract was specifically for the preparatory survey and the setting up of a cannon foundry at Indret on the Loire below Nantes. William was to be paid 120,000 livres, just less than £5,000 in his day, for two years' work to be paid at six-monthly intervals in advance. In comparison with the £50 a year plus house, board and expenses he had been paid as manager at Bersham, this was riches indeed. There was more. All his travelling expenses were to be covered, he was to be paid a lodging and subsistence allowance of 1,000 livres a month and on salary and allowances he would pay no French taxes. If the work was completed in less than two years William was still to be paid the 120,000 livres; if it took longer he would be paid pro-rata for the additional time.

There is an interesting final item in the contract. William reserved the liberty to return to England on two counts: first if the government demanded it, second if his brother died. He undertook, however, to provide a substitute to cover the work in France if this proved necessary. It is an astute point; providing evidence of an awareness of the state of tension between the two governments at the time, and proof for the British government, should it be required, of where the priorities of this established firm of ironmasters lay in the event of war. John would not wish his assets and property in England to be confiscated while his young brother lived royally in France and waited for better times. And if he helped create the terms of this contract he clearly saw his brother William as his successor in England in the event of his own demise. John was approaching 50 at this time, William was 34, and there is still no clear evidence of a breach in their relationship.

It is easy to underestimate William at this point in his life, but if his brother had no part in the negotiation of this contract and did not know the detail of it things may be interpreted very differently. First it would indicate that William in his Bersham years had acquired not only the confidence to take full responsibility for a daunting business commitment in a foreign country, but also the perceptive business acumen of his brother and father. The provision of a substitute to cover his work in France then suggests a temporary return whilst retaining his interest there until he could establish which way the British government would move with regard to his dual responsibilities, or, on the other hand, until he had done what would be necessary to secure his succession in the event of his brother's death. His thinking in either instance may not have been known to John.

William left for France from Bersham at the end of 1776, apparently with his brother's blessing. He took with him a detailed knowledge of all the new technology of the Bersham cannon foundry, much of which during his years as manager he had helped his brother to perfect. It included the latest gun-boring lathe patented by John in 1774, and details of James Watt's new steam engine – the early examples of which were just coming into use, the second version in production at his brother's works at New Willey. An indication of William's competence is the sure and measured way in which he constructed the Indret cannon foundry and brought it into production working in close co-operation with a French engineer called Pierre Toufaire.[14] The Bersham technology built and installed there included gun-boring lathes, reverberatory air furnaces to enable him to use molten metal from scrap in the final casting, with power provided initially, according to a 1788 eyewitness, by waterwheels turned by the tidal ebb and flow of the river, and only later by a Watt-type steam engine:

> Messrs. Espivent had the goodness to attend me on a water expedition, to view the establishment of Mr. Wilkinson for boring cannon, in an island in the Loire below Nantes. Until that well-known English manufacturer arrived the French knew nothing of the art of casting cannon solid, and then boring them. Mr Wilkinson's machinery for boring four cannon is now at work, moved by tide wheels but they have erected a steam engine, with a new apparatus for boring seven more. M. De La Motte, who has the direction of the whole, showed us a model of this engine, about six feet long, five high and four or five broad, which he worked for us by making a small fire under the boiler that is no bigger than a large tea kettle, one of the best machines that I have seen ...[15]

The first cannon were cast at Indret from melted scrap in 1779. William thus fulfilled the terms of his contract, which then appears to have been extended to cover the arrival there of the first French manager of the works: the son of an important family of Lorraine industrialists, François Ignace de Wendel, who was granted a fifteen-year lease from 1780. They were two wealthy young bachelors of about the same age, de Wendel a few years William's senior, and they formed a friendship which would introduce William to some of the most influential people in the country. It is not difficult to imagine the social opportunities open at that time to two such eligible bachelors.

Throughout this period William had kept in contact with his brother John, a valuable kinship which would have given him status in the developing industrial world of France, and he was able to feed him with information which led to new steam engine sales for Boulton and Watt. This enabled John to tender, successfully, for the huge quantity of cast-iron pipework with bores up to 12in and the associated pumping engines that were required in the French government contract to provide a new water supply to Paris from the Seine.[16]

By this time France had aligned with the American states in the American Wars and was thus at war with England, but no government recall had come for William who indeed may have been seen as a valuable source of inside information on enemy affairs. It would have been a difficult course for William to steer though there is no evidence that at this early stage in his French career he acted as a spy.

Much has been made of the Wilkinsons' dealings with France at this time of hostilities between the two countries and extravagant accusations levied against them, from merely treacherous sympathies with an enemy power to charges of actively selling guns to both British and French governments during the war. Huge quantities of iron piping in bores of 3in, 6in and 12in and in lengths of 6ft and 12ft, required for the new Paris water supply, lay on quaysides at both the Severn and Dee ports for long periods waiting for transport across the Channel. This trade had been sanctioned and it is probable that the pipes were mistaken for cannon. Although John made huge profits in the heavy wartime demand for his new improved cannon from the British government, there is no evidence that he was involved in direct sales of cannon to the French at the same time.

However, a measure of the government attitude to continuing trade in a time of hostilities between the two nations is shown in a letter reporting political gossip to John from Joseph Priestley, who was living in London close to the corridors of power as war threatened again during the early 1790s:

> ... That the French do not fear the war is evident enough though it is as evident that they wish to avoid it and are sincerely desirous for our friendship. It is said that the last ambassador, M Manet, was instructed that in case he could not make peace to propose that during the war merchandise should not be captured ...[17]

Certainly new steam engines were exported to France in William's time as a consequence of sales he engineered and approved under passports provided by both governments. It is at first surprising that business activity of this kind continued between two countries at war and was sanctioned by both governments, though it is not difficult to find examples of it in the modern world. The transactions are however, well documented.[18]

The fact remains that during the war William was making improved cannon at Indret for the French with clear access to his brother's works at Bersham. It contributed to the wilder accusations and might have provided a basis for charges of treason, which is perhaps why William had carefully written into his contract with the French the point that he must be at liberty to return to England if his government demanded it.

Far from returning home at this time, however, William took on further work for the French government. He was, after all, well known and well regarded in France following the success of the Indret works; spoke fluent French; and moved easily in the highest social circles. It is significant that when a second cannon foundry was planned by the French Government to be built in Burgundy, known as Louis XVI's New Cannon Foundry, William Wilkinson was asked to undertake the initial survey and bring the works into production. It suggests that William was known to, and approved of, by the king himself. Such credentials would give him immediate authority in his new area of operations and the detail and thoroughness of his initial survey will have enhanced his reputation further.

Recent research has located the original manuscript document in Le Creusot, signed by William and read and approved by his former colleague Pierre Toufaire, the French engineer who helped him set up the Indret works.[19] It is an impressive catalogue of requirements showing precisely how the preparatory work should be staged; where investigations should begin to locate the available ores, the local clays for brick-making and the building timber resources of the region; what labour, materials and buildings would be needed on site at each stage of the process, with cost estimates provided; how substantial savings might be made by good preparation; and how the latest knowledge on iron-making processes from England might be incorporated into the scheme. The survey is presented in sections with sub-headings, is sequential and systematic and concludes with the following:

> ... Slag from Forges. For some years it has been recognised in England that the slag from fineries or bloomeries mixed with the ore for blending in the blast furnace produces great advantages as much in the yield as in the quality of the iron, especially when one is aiming to make wrought iron. I believe it essential to establish this method of working in the furnaces here and for that I request that we acquire slag from the forges of Mesvrin and La Mothe and that it should be brought to the works before the furnaces are lit.

It will be necessary to negotiate with the Masters of these forges in order to secure these slags for the future, for in all likelihood when they see the use that will be made of them, they will be tempted to put up the price or to keep them for themselves.

At Montcenis the 16th October 1781   WILKINSON
Read and approved   P TOUFAIRE

The intention was to use coke as a furnace fuel under François Ignace de Wendel's supervision, as at Indret, and William lists in the survey report how the coking experiments on local coal supplies must be carried out and stocks of coke accumulated before iron production could begin. It is also another pointer toward an aspect of William's character that the management team at Le Creusot was composed of the same professional colleagues he worked with so successfully at Indret. He was paid 50,000 livres a year (just over £2,000 at the 1779 rates of exchange) for the period of the survey which took place between 1779 and 1781, and 72,000 livres a year (£3,000) from the time he was appointed manager of the Le Creusot works in that year until he finally returned to England in 1789.[20] William's total salary in France over a period of nearly thirteen years therefore amounted to almost £30,000 at then values, with allowances, expenses and tax benefits in addition. But the approaching Revolution and the danger to life and property of people in his position threatened this livelihood and broke up his circle of friends. De Wendel fled to Germany. Others were not so fortunate. Arthur Young, travelling in France just before the storm broke, visited both the Indret and the Le Creusot works after William had left:

Nantes is as enflamme in the cause of liberty as any town in France can be; the conversations that I witnessed here prove how great a change is effected in the minds of the French, nor do I believe it will be possible for the present government to last half a century longer, unless the clearest and most dedicated talents be at the helm ...[21]

In Montcenis near Le Creusot on 3 August 1789, three weeks after the storming of the Bastille in Paris and less than 200 miles distant, Arthur Young yet found men speaking approvingly of 'Monsieur Weelkainsong' who they knew to be a brother-in-law of Dr Priestley and, therefore, a friend of mankind, and that he taught them to bore cannon in order to give liberty to America. In that context there is, however, a note of warning in Young's description of the Le Creusot works: 'The establishment is very considerable; there are from 500 to 600 men employed, besides colliers; five steam engines are erected for giving the blasts and for boring; and a new one building ...'[22]

When he returned to England William had clearly become accustomed to wealth and status, and had proved that he was capable of bringing into production and successfully managing ironworks to match those of his brother. He would be looking for an opportunity within his brother's empire appropriate to his French standing and success. Almost immediately friction developed between them.

The question arises, then, as to whether John saw William's return as a threat to his own position, and how far the attitude and expectations of William provoked this. Undoubtedly, the William who returned would be a prouder, more assured and more sophisticated man than the William who had left thirteen years before, but it would be difficult for John to treat the new William as an equal. Of itself that would be enough for differences of opinion immediately and tension, ultimately friction. Their sister Mary (from evidence in the Priestley letters) is also known to have quarrelled with William at this time. Reporting her death in America in 1796 her husband, Dr Joseph Priestley, wrote to John:

> ... She always warmly took your part and would never believe your father's account of your using him ill. To your brother William she had the affection of a mother but his behaviour to her on his return from France shocked her in such a manner as I cannot describe and she never recovered it ...[23]

It is likely, therefore, that there was arrogance and insensitive braggadocio in the attitude and posture of William on his return to England, which family and friends other than John found difficult to accept.

There is, however, evidence of a failure by John to pay William his due share of profits as a partner in the Bersham works throughout the period of his stay in France.[24] John might have thought that William, with his huge French salary, did not need the money, but had acknowledged somewhat ungraciously that profits were owing to him from 1777, when William raised the matter at a meeting in Belgium in 1782. William had asked for a written acknowledgement of this and confirmation of his ⅛ share in the Bersham works, and this seems to have been provided. Since nothing had been paid to William up to the time he returned temporarily in 1784, he went through the Bersham accounts and found many irregularities, which he took up with John who then agreed that £800 should be taken from the Bersham profits and paid into their private accounts in the ratio 7:1. William said no money had been paid to him up to the time he left France for good in 1789.

An earlier return from France by William is recorded by Anne Watt in a letter to her husband in August 1786, when William dined with her before setting off for Castle Head.[25] Indeed it seems clear that during the three years between 1786 and 1789 he moved frequently between France and the Wilkinson properties in England. It was during this period that he made a detailed inventory of the various Ironworks in France and of the tonnage and type of iron made there.[26] The document dated 5 February 1787 is not signed and is not addressed to any person or concern, and it might have been simply for William's own reference on his return to England, which he clearly saw as imminent, his prospects becoming uncertain as the Revolution neared. Such detailed information would be extremely useful to any future employer.

William took the opportunity during these return visits to examine further his brother's books of accounts, which in the case of the Bersham and the Snedshill Works was his right as a shareholder. The complaints of irregularities he made to his brother clearly fell on stony ground and their differences then festered until his final return.

At that point William offered to sell his Bersham interest to his brother 'for a sum greatly below the value thereof'.[27] John's response, in a reply dated 30 October 1789, was dismissive: 'have come to a Resolution not to sell or buy. I should decline the latter did you offer it for the cons'n of 5/– a sum in the Law used to convey a gift …'[28] Their disagreements then steadily escalated into the bitter dispute which ended in the litigation of 1795 and 1796.

10

DISAGREEMENT, DISPUTE AND LITIGATION

As the dispute with his brother deteriorated into bitter recriminations on each side, John must have realised that the days were numbered for the New Bersham Company as a hugely profitable enterprise effectively under his sole control, and he began to look around for alternatives. With his preference for locating in one place as many of the processes attendant upon iron-making as possible, he was seeking a small estate in the area with its own mineral resources. In 1791 he bought such an estate at Hadley, East Shropshire, though perhaps at that time with a view to replacing his New Willey works as his lease there ended, and in fact he did not immediately build a furnace and ironworks there.

In June 1792 he bought the Brymbo estate in Denbighshire. It came with an impressive Mansion House designed by Inigo Jones and lands which provided him not only with a source of iron ore but also with coal deposits.[1] He clearly intended to develop another large industrial complex here just 5 miles to the north of Bersham. The estate cost £14,000 and he had to call in credit to pay for it,[2] and an indication of a waning confidence in his affairs began to show in a new attitude by tradesmen to his money tokens, by then circulating freely in local markets.[3]

The worsening dispute with his brother can be traced in the letters during this period between John and his young protégé from Kendal: then

John Wilkinson remained a commanding figure late into his life.

his Clerk of Works at Bersham, Gilbert Gilpin. With an obvious regard for both brothers, yet employed by John, Gilbert's position was a difficult one and his letters make clear that in attempts to manage the extreme positions of each of them he lost the confidence of both. As the dispute progressed his attitude to John, his employer, became less sympathetic while continuing to protest his loyalty and support, though it was William who would eventually threaten him with litigation for what he perceived to be an obstructive interference to his interests. Clearly, both brothers made him listen to their own side of the case so that Gilbert was perhaps better informed about their dispute than anyone else. In what amounted to his resignation letter, he made one last determined attempt to mediate:

Sir, As near as I can recollect the last time that W.W. was at the works was the 15th of last month. He did come into the counting hse but not attempt to look into the books. He was in the counting house when your letter ordering that he was not to see the books in future arrived, & I in course showed it to him. He intimated that such refusal was then of no consequence because he had paid sufficient attention to those matters for some years past; & that though you had denied him access to the books (which you had no right to do even if he was possessed of only a hundredth part of the concern) he would have all the information he wanted brought into court.

Untill very lately he used to come to the work as usual, & in all the conversations which we have had on the subjects of the dispute which exists between you and him he declared, that as you and him could not agree to settle it yourselves, he had proposed some years ago to refer the whole to Dr Priestley, but that you declined it, and wrote a note which is now in his possession (a copy of which will form a part of the amendment to the Bersham Bill) saying that you could not agree to have the business settled in that manner untill you had sent J. Priestley to the continent to learn what engagements he had entered into there that might be binding upon you. He had repeatedly requested that the matter might be settled by arbitration but that you always evaded it & which would be proved by your own hand writing at a proper time. That seeing no other method of procuring a settlement, and you having particularly advised him to [?] he had been under the disagreeable necessity of endeavouring to obtain it by the laws of his country.

His reason for not answering Mr Harper's letters was, that as you seemed determined to have no correspondence with him but through some organ of the law you should have an answer in that way at a proper time. Had you wrote to him yourself, or were you to do so now, I am persuaded you would have an answer immediately; and I am pretty certain that he would have no manner of objection even now to have the matter referred to your mutual friends ...

I have come to a determination to try my fortune in the new world. The time of my departure I wish to make as convenient as possible to you, & could I get off with the ships that sail in the latter end of April or the beginning of May I should be perfectly satisfied.

I am sir, Yr. hble ser.

G.G.[4]

Gilbert's attempt at reconciliation was to no avail. He was summoned by John to Castle Head in April 1794, taking the Bersham books and papers to study William's Bills of Complaint and to help provide some answers.

This conference will have contributed to the strategy John evolved to deal with William's complaints; in which he sold, or tried to sell, his interest in the Bersham works to three of his relations in exchange for a bond from each of them.[5] They were Richard Watson, grandson of his father's eldest brother and by then his agent at Castle Head; Thomas Jones, son of his sister Sarah and closely involved with the Bersham management up to this time; and William Johnston, son of his mother's brother and then manager of the Bradley Ironworks.[6] From what Gilbert Gilpin says in a later letter, it seems unlikely that any money actually changed hands at the time and in fact the transaction in the later litigation would have put the affairs of all three of them at risk.[7] John continued to refuse to deal with William directly, however, referring to him as 'a self created & violent turbulent Person bent upon Mischief'. In the summer of 1795 the hearings began.

William's Bill of Complaint to Chancery was detailed and carefully presented. He was owed a ⅛ share of the profits in the Bersham Company from 1777 to that date, and a ⅛ share of the profits in the Snedshill Works from 1778 to the end of the lease in 1793. None of this money had been credited to him and his attempts to establish his claim by reference to the company records had now been frustrated by his brother's denying him access to the books.

However, his earlier examination of the books had shown that the Bersham works were very profitable and that the recent lead smelting furnaces there had increased profits; that John continued to shift money between his various businesses at his own financial convenience without proper accounting; that he had used £16,000 of the Bersham profits to increase the company capital without reference to William who, as a partner, had been billed for ⅛ of this amount when he should have been paid it; that he had concealed transfers out of the Bersham accounts to his other businesses when the Bersham profits were high, particularly to the London Lead Company at Rotherhithe, thus avoiding tax and depriving William of his dues; and that clerks of works at all his business premises were now instructed not to allow William access to their books.

In his reply, predominantly with reference to William's ⅛ share in the Snedshill company which he acknowledged, John dismissed this 'swaggering epistle' as clear evidence that William did not understand the detail of company accounting procedures. Since the Snedshill Company affairs were not finally settled and a closing balance sheet was not yet available to enable profits to be paid out, John said his claims were premature and unnecessary. Moreover, William had deliberately tried to create confusion between the two companies, 'the said Furnace at Snedshill being always considered as part of and appendage to the said concern at Bersham'.

This appears reasonable until John 'humbly hopes' the court will not require his 'General Books of Account … which contain much irrelevant matter', though he agrees to produce his weighing books, which show weights 'IN' of raw materials and weights 'OUT' of iron deliveries from the furnace.

John then denies to the court any recollection of writing any of the letters attributed to him, of which William has produced copies, and of the verbatim quotes used by William in support of his claim. He answers William's claims of financial malpractice point by point, acknowledging his ⅛ share in the Bersham Company but denying any fraudulent accounting, claiming again William had not understood the accounting procedures. The impression remained at the end of the evidence in this claim and counter-claim, however, that Bersham had been hugely profitable and that nothing had been paid to William for his ⅛ share between 1777 and 1794.

Throughout the period of litigation John tried to frustrate the court directions again and again by delays in responding and requests for more time. The eventual outcome was an agreement from each side to go to arbitration. Since this had been urged from William's side before the costly litigation began, it must be that at this late stage John could see a decision of the court going against him. Four arbitrators were agreed: William Fawcett, Thomas Bennion, William Robertson and William Reynolds. Their award, though largely in favour of William, did try to strike a balance.[8]

The costs of the arbitration were to be paid out of the Bersham Company business. John objected to this since the value of their holdings in the business were 7:1 in his favour, and this meant that John was effectively paying most of William's arbitration costs. Next, William's claim for back payment of monies due to him out of the partnership were accepted. He was awarded £8,850 to be paid in two equal instalments direct to him from the partnership bank, Messrs Eyton, Reynolds & Wilkinson, at six-monthly intervals over the following year. Each side was to bear its own litigation costs at law and in equity, including the costs of getting their respective bills and claims against each other dismissed by the end of the next term (Trinity). Finally, William was to be available as a future signatory to any outstanding partnership transactions that required it, with any costs involved to be paid by John. The award was dated 2 May 1796.

In a letter to Boulton and Watt enclosing these details, William said it was 'the first Meeting of the Arbitrators', and that it indicated 'their Determination to sell the Works and dissolve the Partnership. And in the Mean Time to convert the divers Effects into Cash'.[9] Since this is what ultimately happened, it seems that at least one further meeting was held. William's account in the same letter of what happened at the meeting of 2 May 1796 is revealing:

> ... The Meeting was very amicable as far as the Arbitrators were concerned but My Brother would not join us and he sat in an Adjoining Room where he had everything conveyed to him by Mr Reynolds and his Meat was sent him from our Table so that all heat was prevented and the Meeting was as pleasant as a Business of this Nature could be ...[10]

It did not end there. William had to go back to the court for a further order against John's non-compliance with the arbitration award before the matter was resolved, and there are interesting references in other letters of the mid-1790s to this bitter quarrel between the brothers, which endured until they died.[11]

Gilbert Gilpin, whose sympathies in the dispute increasingly lay with William, left John's employ shortly after judgement had been given, with neither his permission nor his blessing. It was a cause of anxiety to Gilbert over the next few years as he tried to find another position among rumours that John intended to take action against him. Gilbert's rambling gossipy letters, mostly to William Wilkinson, may be neither entirely reliable nor impartial but they reflect a close knowledge of the late eighteenth-century iron trade and the interconnected personalities involved:

> ... In respect to your brother's assertion ... He may perhaps still have it in idea to prevent any person, or persons, who do not keep upon good terms with him from doing anything in the iron line in England, France or America ...
>
> But the most material matter which gave rise to my conjecture was the report of a Mr Wilson of Sheffield who I met with in the coffee room at the Shakespear in Birmingham about three weeks ago. On hearing that I had been an agent of Mr W he informed me that on coming through Wrexham he heard two men discoursing on the subject of a Mr Gilpin who had lately left the employment of Mr W against his consent, & that it was the intention of the said Mr W to do him all the injury in his power ...[12]

It is perhaps not surprising, therefore, that Gilbert rejoiced whenever he heard anything to the discredit of his former master and reported it to his correspondents at every opportunity. The subject of the following conversation is Mr Reynolds, who was the William Reynolds of the arbitration board and thought to be sympathetic to John rather than William:

> [I] then mentioned the offer of Mr Fawcett, which he recommended me by all means to accept, & then mentioned the probability of an opposition from J.W. in such case. This he made a laugh at & said, 'What does any of his oppositions effect? He is now selling his rods at 16/- & we ours at 22/-' ...[13]

In the summer of 1794, Dr Joseph Priestley and his family had emigrated to America. He was financially beholden to both brothers-in-law, but more so to John who had supported him for many years. Throughout the following year he received at least three reports from Richard Watson about the progress of the lawsuit, clearly on John's instruction, as well as sets of accounts and complaints from both brothers.[14]

His sympathies clearly lay with John, applauding what he sees as 'the judgement and still more the excellent temper ... of your several attempts to bring the dispute to an amicable reference'. He had tried to reason with William from whom he had received 'only short letters on the subject ... from which it is impossible to form any clear idea of the nature of his complaints' – all to no avail. He then urged them to find a mediator who could lead them back from recourse to litigation and the ruin that he feared would ensue. Financial ruin for the Wilkinsons meant, of course, an end to the financial subsidies he received from them himself and he clearly felt very exposed on this count as a pioneer settler in America. He mentions Samuel More as a possible mediator but he by then was seriously ill and not expected to live.

There is an interesting glimpse into a significant moment in the dispute in one of the Priestley letters.[15] After words of solace and reassurance to John, Priestley says:

> ... The same ship that brought Mr Watson's letter of June 26th brought me one from your brother of May 26th, a copy of my answer to which I enclose though I wish you would not give him any intimation that I have done so. What he says (and what you want to know) concerning your proposal to him to take back [?] is as follows: 'At the end of six days when he found things to go against him he decided I would take the whole works at Bersham and Bradley into his – he must have meant my – hands and that he would give up all the direction and give me a bond not to interfere and would even

withdraw his name from the concern. To this indignant offer I replied I would sooner put my hand into a fire than accede.' ...

Priestley's anxieties about the danger to his financial support from John if the dispute came to trial were justified. News of the settlement, which he clearly thought was too heavily in William's favour and unfair to John, was quickly followed by a letter from Richard Watson declaring that Priestley was more than £6,000 in debt to John, this reckoning almost certainly a result of the legal process and the consequent listing of John's assets. With the spectre of further litigation looming he looked for reassurance from John about Richard Watson's letter and urged him to accept the result with resignation and sue for peace. It did not happen.

Either in anticipation, or as a consequence, of the arbitration award John first tried to dispose of his shareholding to his relations as described previously, clearly with the intention of retaining ultimate control. William would have none of it and his dismissal of this plan appears to have been one of the issues the arbitrators had to settle.

A sale of Bersham, to take place at the end of November 1795, was arranged by the arbitrators. In the weeks beforehand William appeared confident of his position and increasingly disposed to bid for the Bersham works himself.[16] He secretly involved himself in a plan with James Watt Jnr to smuggle into the works a man 'who you are sure my brother or his satellites did not know' to make an inventory of the goods and machinery on site, the better to enable William to make his bid. One can imagine all machinery stopped at the great ironworks at this time; the furnaces blown out, and the place largely deserted and silent, locked up and under guard.

In the event, William did purchase the Bersham ironworks and so became responsible once more for the works where he began his career as an ironmaster. There is sadness, disappointment and even bitterness in his brother's letter to Matthew Boulton the following week:

8th December 1795
Dear Sir,

Your favour of the 29th ult. was delivered me by your son but my engagements during the sale of Bersham and the things attendant upon it, which put that place once more into the hands of its old Possessor, took so much of my thoughts that I had it not in my power to answer it on his return ...

There is now an end of all connection between W.W. and me, except in closing the Accounts which will be done prior to the meeting of the Arbitrators in April next ...[17]

John by this time had his new works at Brymbo in production. His policy before and after the sale to William would have been to claim the major part of the Bersham moveable goods and machinery as his own and transfer them the 5 miles to Brymbo, as indeed he was permitted to do under the arbitration award. It would nonetheless be a further cause of continuing bitterness between them.

Both brothers appeared anxious to retain the goodwill of Boulton and Watt, who were now old men like themselves. John's approach, in sincere and measured language, was to reassure them as creditors of the New Bersham Company up to the time of the sale, with a concern to see their unpaid bills transferred out of the arbitration process direct to him for payment. He referred to the long-standing trust and friendship between them threatened in this dispute with his brother, and hoped that 'the Unity of the Trinity which once subsisted may probably again take place'. He reassured them of his earnest wish to resolve all grievances between them and received in turn from them the wish 'to preserve that friendship, peace and harmony' of earlier days.

William had a different approach. As soon as the Bersham works were sold he began the invidious process of turning opinion against his brother, particularly through his contact with the Boulton and Watt operations. He felt he had good allies there in Matthew Robinson Boulton and James Watt Jnr, who had been deputed by their parents to find, up and down the country, the new Watt steam engines that had been erected without licences and were known as pirate engines.

Because of the delays in obtaining the licences and the consequent frustration to intended purchasers, John had been advising his friends, and notably James Stockdale of Cark, to proceed independently of Boulton and Watt.[18] He then provided the iron cylinders and parts for the engines either himself or through his contacts direct to the purchaser, with the services of an engine erector thrown in but without the knowledge of Boulton and Watt. In granting a licence to use their engines Boulton and Watt contracted to receive premiums at half-yearly intervals for the remaining duration of their patent, calculated on the basis of a one-third fuel saving on the old Newcomen engine. These pirate engines, therefore, cheated them of substantial revenue since no premiums were paid.

Soon after joining the company the two sons, with the help of William Wilkinson and other spies, were commissioned to find the pirate engines and by recourse to litigation (though often the threat of it was sufficient) secure the necessary premiums for the company and backdate them.[19] William exploited these opportunities to discredit his brother in the mid-1790s and even succeeded in driving a wedge between John and his old friend and agent, James Stockdale of Cark. It came about in a curious way.

On New Year's Day 1791 William, then 47 years old, had married for the first time. His wife was a widow, Mrs Elizabeth Kirkes, then living in Liverpool and a daughter of James Stockdale of Cark. This meant that as the dispute with his brother grew in strength and bitterness, William had close and regular access, as an intimate within the Stockdale family, to one of his brother's oldest friends. It is known that he provided information to Boulton and Watt about his brother's pirate engines, but then came the curious twist in the story.

Boulton and Watt found a clear breach of their patent in the Cark engine and took action against the company. Although he might not have known until late that the Cark engine was a pirate engine, William had clearly decided to keep quiet about it anyway. He then realised the difficulties of his position. On the one hand he had supported Boulton and Watt in their campaign against these engines and was particularly friendly with the two sons who were leading the prosecutions. On the other his own father-in-law, a well-respected figure in the industrial and commercial affairs of the day and an old friend of Boulton and Watt Snr, now stood accused.

When it was clear that Boulton and Watt intended to proceed against Thackeray Stockdale & Co., William moved quickly. First he summoned his father-in-law to his home at Court near Wrexham: 'I am <u>particularly desirous</u> of seeing you here upon a Business which I cannot write you upon and which concerns you. I think I can be of service to you therein and the sooner you come here the better ...'[20]

The letter has an urgent, even imperious, tone. William's intention was to set up a face-to-face meeting with James Watt Jnr to try to settle the affair quietly and so protect his father-in-law. Perhaps surprisingly there was massive goodwill towards their old friend from the Boulton and Watt camp, who chose to proceed against Thackeray and Bradley, the maker of the pirate engine parts, without naming James Stockdale in the litigation. This decision is confirmed in a letter from James Watt Jnr shortly after the meeting.[21] A month later his father was also writing to thank William for his friendly interference: 'We are very much concerned that Mr Stockdale should be implicated anyways in this matter ... no unfriendly steps should be taken against him or the other innocent members of the Cark Company.'[22]

William was able to be of further service to his father-in-law when Thackeray, after due procrastination, finally agreed to accept Boulton and Watt's nomination of him as arbitrator to establish what money was due in settlement. He managed to reduce their claim from £762 10s to £550, which Boulton and Watt accepted 'with perfect satisfaction'.

It would be helpful to know how William and his father-in-law related to one another at this time. The slightly high-handed tone of William's letter summoning his father-in-law to Wrexham could suggest William's dominance and an impatience and embarrassment through his connection. On the other hand, his wife had just produced their first baby girl and it might have been a kindly intervention to support a well-respected old man who was failing. Certainly it will have given William some influence with James Stockdale, as was confirmed in a letter about this time to Matthew Boulton about the approaching arbitrators' meeting in his own dispute with his brother:

> Dear Sir,
>
> I am desired by Mr James Stockdale of Cark to acquaint you that He and my Brother are about getting quit of each other and settling their Transactions in Business. They have ... copper at Stourport one half of which ... Mr Stockdale has ordered to be delivered to your care ... My Brother wants to postpone the Settlement of our Disputes until another year and objects to let me have access to the Books. As this is expressly provided for by the rule of Court I shall see Mr Reynolds on the way and in Case I cannot have that right without an Appeal to the Court of Kings Bench that must take place the next Term. I am more and more convinced that Nothing but Compulsion will ever induce JW to settle our disputes however I am not afraid of him ...[23]

The arbitrators' meeting listed for the middle of April 1796 to close the Bersham books was further delayed at John's request but with William's concurrence, which means he, too, wasn't quite ready for the denouement. He had been busy in the spring and early summer breaking up the old loyalties to his brother at the Bersham works by getting rid of key figures like Abram Storey, clearly a man of some authority there; Thomas Matthews, a borer and turner at Bersham for twenty-six years; and one Kendrick who seemed to be completing orders that had been in hand at the time of the sale. Most of these men he tried to place in the Boulton and Watt empire. By July, Abram Storey was already there:

> ... A. Storey by writing to Kendrick might have the choice of any men at Bersham as they are chiefly leaving it ... [And then a new attempt to discredit his brother:] ... Abram's successor was at Brymbo ten days but as he never was so sober as to get up and dress himself he has been discharged and is to be replaced by a family from South Wales whom Crawshay has discharged for being drunk and insolent. JW will have no other iron than that of Brymbo which on being remelted into guns is so hard as not to be bor'd, but none else will be purchased ...[24]

It seems likely that by this time William had decided to sell the Bersham ironworks and would have had Boulton and Watt in mind as prospective purchasers. He was dealing with a canny businessman of course, and Matthew Boulton would have had no scruples about accepting any early benefits of a potential transaction without committing himself. William, meantime, clearly wished to play down both the Bersham ironworks and his brother's reduced empire as any kind of competition.

By this time the arbitrators had pronounced on the complications of the Bersham accounts, with the financial settlement between the brothers in the final closing of the books very substantially in favour of William. He was well pleased with himself:

> ... JW is no Economist of harsh terms in respect to the Arbitrators whom he blames as having acted very unjustly, and sent for Mr Fawcett since you saw him at Soho to get him to sign a Voluminous Number of Observations drawn up by JW and Watson, all tending to accuse the Arbitrators himself and Gilpin of partiality for me in the settlement. Fawcett went home much displeased at JW's ideas of his consistency as he informed me that he could not do it if he had given him £20,000. JW stated the necessity of it saying in the present state it appeared he had acted unjustly and that he should be dishonoured amongst his acquaintance to whom he wished to shew Mr Fawcett's Approbation of his Statement of the case between us.
>
> This I hope will amuse you as well as that he sends the Creditors of the late concern to Me for Payment. I have wrote Weston and expect we shall be obliged to move the Court for an Attachment against him for Contempt as the Time for his objecting to the Award is past and now he objects to the Performance [of] it in certain respects. He won't pay the Expenses of Fawcett or the Travell Charges of the Arbitrators which he is ordered to do. Watson and he are like to quarrel as he won't give him anything for his Time or pay him for his Living with him for he has not got the better of Me. His conduct is at once Mean and laughable ...[25]

It is clear from William's letter that John was bitter about the arbitrators' decisions and had begun a process of attrition; wasting William's resources by forcing him repeatedly back to the court for new and confirmatory judgements. The case of the Maas-y-Fynnon lead mine is a good illustration.

John had been the principal for thirteen years in this concern, which had provided lead ore prior to the sale of Bersham to the very profitable lead furnaces he had established there. On 29 October 1796, following the arbitrators' settlement in the spring and what was clearly a very stormy general meeting with opposing sides present held that day at the Eagle Inn, Wrexham, John posted handbills in the area of the mine dismissing Robert Burton as mine agent and appointing a Mr William Jones of Pwlygo in his place.[26] John's supporters included Richard Watson and Thomas Jones, his nephew, who was alleged to have power of attorney for the votes of Joseph Thackeray and Benjamin Satterthwaite, principal defendants in the Cark engine litigation.

William riposted immediately with another handbill:

> … The Public are hereby informed that the above Hand-bill is a total misrepresentation of facts, as John Wilkinson, Thomas Jones, John Jones, Richard Watson, Joseph Thackeray, and Benjamin Satterthwaite are no longer Partners in the Concern, having severally forfeited their shares therein for neglect in paying their quotas of a call to reimburse the expenses incurred in prosecuting the work, being first thereunto required by notice in the London Gazette of the 17th of September last …
>
> The Public are also informed that at the meeting alluded to in the Hand-bill, Mr Burton was continued the Agent and Treasurer of the Concern by a majority in value of the <u>real</u> Partners present; and it was the opinion of such Partners that it was unadvisable in them to admit Mr John Wilkinson director of the Concern, he having to the present time refused to render satisfactory Accounts during the space of <u>thirteen</u> years he was entrusted with the management thereof, or even to permit the Partners to inspect and investigate the same, altho' required by several of their resolutions so to do; and for which legal proceedings are instituting against him. We shall now be silent and wait the event of a legal investigation …[27]

John's newly appointed acting agent to the mine, William Jones, posted yet another handbill the following day saying this was all nonsense.[28] His principals, Jones said, paid their share of the costs in full, the account books showed this and would be kept in a safe place until Robert Burton cleared up his own accounts of expenses in the mine, 'which there appears no other Means of making him do but by Legal Proceedings'. The stage was, therefore, set for yet another expensive legal battle which would waste the energies and resources of both sides.

John's differences with Boulton and Watt were only partly due to the problems of disentangling his accounts with them at Bersham in the escalation of his dispute with William – who now clearly wished to sour their long-standing business relationship if he could. With weasel words of derogation here and there in his own letters to Boulton and Watt, and particularly to their sons, he took every opportunity to do so and made no secret of it. John was certainly aware of what was happening but hoped the relationship with his old business associates would be strong enough to survive such tactics. As his dispute with William reached a climax he wrote to Matthew Boulton:

> ... When the disagreeable business which at present takes up too much of my time is ended I hope to have an opportunity of seeing you, and am persuaded that all apparent clashing of sentiments and interests can easily be rectified, and that you will be convinced I have at all times been much more your friend than the person who has taken such pains to injure me in your opinion, and to blow up if possible a flame between us. I trust however he will always fail in the attempt, and that when we meet a plan may be adopted by which our interests may be more closely than ever united, both in the engine and foundry business ...[29]

Their long association had never been completely harmonious and survived largely because it was mutually useful and brought good profits to both sides. They were never friends in the way that John and Samuel More were friends, and although they had social and family contact at each other's homes theirs was fundamentally a business relationship.

From its earliest days, within the so-called trinity there were tensions. Watt, the engineer and inventor, was never happy with the business pressure of profit and loss and supply and demand, which of course was Boulton's strength, though he had a good understanding of iron-making and furnace systems, too. John related strongly to the practical side of Watt, whom he also liked as a man. Boulton he respected for his entrepreneurial and marketing skills with an acknowledged status and a ruthlessness as a businessman to match his own. Perhaps because of this there was always a reserve between the two. Watt's second wife, Anne, recorded some early signs of disagreement between them all in the mid-1780s:

> ... I am sorry to hear so dismal an account of the mines and am afraid Mr B is not the man to reform them. He talks too much. Mr Wilkinson is the man to go among them. Perhaps Mr B is like some doctors who strive to make the case they are to undertake to cure a desperate one that their fame may be the greater ...[30]

The letter suggests that the Watts by this time had already shared some reservations about Matthew Boulton, and they continued to do so:

> ... You say Mr B has great hopes of the year 86 tho I like you don't build much on his airy views ... [he] is not to be trusted with money for had he millions he would find ways to spend it ...[31]

And a week later:

> ... I can not say but I am very sorry to hear Dr Withering is going to the opposite side of the town from us & in my rage for the loss of so valuable a neighbour abused Mr B in my own mind worse than a dog for him I look on as the sole cause of it don't you think ... I am really very angry and think he has acted a very unfriendly part. Had we been a Lord or Duke he would have strained every nerve to have served us. I now almost hate the man ...[32]

By the following month their mistrust of Matthew Boulton had become a serious issue. Watt had obviously shared confidences with his wife about Boulton's entrepreneurial activities. Knowing her husband to be a worrier and a depressive she tried to steady him:

> ...You must lay down some new plan with regard to your conduct with Mr B but that matter we will talk over when we meet if you will but camly [sic] talk on a subject that so nearly concerns your peace – for what is all the world without contentment – & he seems to be hurrying on from one scheme to another and will forever drain you of all your cash ...[33]

Since the Watts were about to meet at Castle Head for an extended stay as John's guests, it must follow that the chances were high of their joint discussions of the conduct of Matthew Boulton, as referred to in this letter.

It was about this time, as Boulton and Watt began to realise their investments in Cornwall were not going to yield the returns they had hoped for and their cash flow was in serious trouble, that Watt again raised with Matthew Boulton the question of bringing in John as a partner. It did not happen and Boulton must, therefore, have been against it. In spite of John's central importance to the production of the new steam engines, by supplying cylinders and fittings for this expanding Boulton and Watt businesss, he received no share in their profits from these sales and depended for his own profits on the sale of the parts to the company. A measure of the difficulties they were under in this process on account of their different priorities is indicated by Watt himself:

> ... Whenever an opportunity occurred of getting an order for a considerable set of castings Mr. W'n was earnest with us to come to conclusion with the customer whether it appeared to us to be for our interest or not – this was the case with Jary from whom Mr Wil'n had a good order and was paid, we had much trouble & a small premium which was never paid – if it had was not compensation for our trouble ...[34]

When delays built up in the orders for engines because Boulton insisted on having an engine contract with premiums to be paid, agreed and signed before he would allow it to be erected, John became increasingly frustrated. It meant that he often anticipated the process and obtained the drawings and specifications from Watt before the contract was agreed and had the parts ready and waiting because he realised that further delay would lead to loss of business. The whole thing often worked to his detriment. There was increased risk involved because the intending purchaser might withdraw; his own time margins between outlay and profit were increased no matter how efficient his production systems; and cylinders and their related parts frequently lay about at his works awaiting the contract before transport to their destination could be arranged. Sometimes the parts went missing.

It was as a consequence of these frustrations that John obtained reluctant authority from Boulton and Watt to erect himself certain engines for his own use, which then led to his involvement in the pirate engine scandals by short-circuiting the time-consuming Boulton and Watt process for his friends. In spite of these serious differences of perspective and of practice the Boulton-Watt-Wilkinson trinity continued to work through the late 1770s and '80s because they continued to make money and each realised that without the others this lucrative business would be threatened. Then came the brothers' dispute at Bersham.

In spite of John's expressly declared wish to resolve their differences and retrieve the old mutually useful harmony, Boulton and Watt were polite and concurred where they could but maintained a distance. A number of issues were relevant. First, John, with his loss of Bersham and with the New Willey lease ending, was clearly not the power in the land that he had been. Second, with William's connivance there were skilled workmen available to Boulton and Watt with the running-down of the Bersham works. Third, Watt's patent was due to expire in 1800 and with John involved in a host of pirate engines, and with their sons' increasing success in obtaining the back premiums for them through litigation, there was money to be realised to replace any shortfall in profits from new engines. That, of course, could only be to the detriment of a continuing business relationship with John. A number of references illustrate this tentative balance between them in the period around the arbitrators' award in the spring of 1796.

Towards the end of that year Boulton and Watt became heavily involved in London in the defence of their patent, against what they claimed had been a breach of it in a new steam engine then being advertised by Jabeth Hornblower. The Hornblowers had previously canvassed support for their engine among the successful industrial magnates of the day, one of whom was John. Matthew

Boulton knew this and, following John's overtures of continued friendship, wrote to him. He first extended an invitation to his home at Soho, though knowing that John had not been to Birmingham since the Birmingham Riots of 1791 destroyed the property and livelihood of his brother-in-law Joseph Priestley. Boulton then said: 'I should also esteem it a mark of your friendship if you would send me the original letter which you recvd from Hornblower inviting you to join him and other such like friends in their opposition to our Patent.'[35]

John's reply was masterly. He wrote from Brymbo:

Dear Sir,

The letter in question ... being at Broseley with other papers in the B&W's bundles and locked up, until I go there in the course of next month the injunction it is under will remain; in the meantime I must consider the propriety of giving up a letter of that nature, more especially if the request I now make be not complied with, viz: that B&W send me the originals of W.W.'s letters to their house or to any of the parties or agents concerned for them ...[36]

No reply has survived but it seems likely such a letter was met with a deafening silence from Boulton and Watt, and that the exchange did nothing to further their continued expressions of friendship.

James Watt Snr had already drawn up a memorandum on their long-term relations with John following the company's listing, probably by the sons, 'of Engines made by John Wilkinson at Bersham for his own and other persons uses, with and without our consent'.[37] The list itself is informative. It covers the period from the very first of Watt's new steam engines in 1776 to the summer of 1795 and is divided into two parts.

The first period, from 1776 to 1784, lists ten engines made by John under agreed modifications to the normal Boulton and Watt contract and premiums. For the first three engines erected by John for his own use no premiums were charged. This was of course in the halcyon days of 1776 and 1777 when there was excitement in the air and an urgency to have the first engines up and running. John's part in this process, beyond that of supplying cylinders and cast-iron parts, was as the practical experimental engineer who erected the engines according to instructions and reported back with suggestions for modifications. It was a key role and Boulton and Watt clearly recognised it by giving up their premiums on John's first engines. The list refers to a letter of 17 July 1777 authorising this.

The next seven engines either had no written contractual agreement, or a modified one to John's advantage, and included five for his own works: two for Snedshill, two for Bradley and one for the lead mine at Maas-y-Fynnon. The remaining two were for France. The Boulton and Watt comments included in the list indicate that John was central in all the modified contractual arrangements for payment, but having agreed the details then paid nothing.

The second list of engines built between 1787 and 1795 is headed 'All the subsequent Engines have been erected without our knowledge or licence'. There are thirty-five engines in this category with three others crossed out. The Cark engine is not included. Boulton and Watt's notes indicate John was involved in all the engines. An astonishing fourteen over the years right up to 1795 were for Bradley, which confirms the size and dominance of this ironworks during the period. The list also establishes that John believed his role as experimental engineer, key supplier of parts and promoter and marketing agent entitled him, in the absence of a partnership in the company, to other financial benefits and concessions and he was determined to take them. Unfortunately, Boulton and Watt did not confront him about this until firm patterns had been established. The reason why is in the elder James Watt's memorandum:

... If it be asked why we did not sooner bring Mr. W'n to a settlement of accounts or take more decisive measures with him in respect to the Piracies he was always committing upon us, the answer is that we know him to be of a vindictive temper (although friendly to us where his own immediate interest was not concerned to the contrary & his friendship we have reason to think was sincere) & we knew he could do us much injury were he so disposed. His friendship for us met with a reciprocation on our side, which we have shown by our attention to his interest always when in our power. He in general executed our work well and there was no period, since the commencement of our connexion, in which a breach with him would not materially have deranged the order of our business, therefore we were not disposed to pursue any measures which might have interrupted the friendship which subsisted with him hoping always that some period he might have been disposed to have made us some reasonable allowances for the use he made of our engines ...

Mr. W's demand of fixt sum for all the Engines he might please to erect, shows the arbitrary & uncontracktable nature of a man who wished for perfect liberty to make as free with his neighbours property as suited him. We cannot help however agreeing with him that these unsettled accounts were the seed of a plant of a very mischievous nature which has under his cultivation brought forth poisonous fruits insouciant at that shrine ...[38]

There is some intemperate language in these notes – an indication of Watt's mood of impatience and exasperation at the time – but a picture also emerges of John as a lateral thinker playing with ideas. Watt talks of the 'purpetual importunity of Mr Wilkinson ... to erect such engines as he might want for his own use', and of John 'perpetually scheming new works ... as if he had been doing us a favour', when in fact it 'cost us much meditation and labour in the contrivance'. That is an important testimonial to John coming from Watt and at this particular time.

In his notes Watt does point to other causes of the rift between them, which provide important incidental information on their personal relations. He mentions the sudden loss of much regular social contact following the Birmingham Riots of 1791, after which John refused to visit their houses any more, depriving them of 'the opportunities of social fellowship ... in which grievance on both sides were discussed & if [not] settled, at least were not placed to the account of want of cordiality'. There is also an interesting reference to John's closing of the Bersham works during the litigation and before the arbitrators' judgement:

> ... his unfortunate contest with his brother made him suspicious of us and in a fit of rashness, probably with an intention to be revenged on both parties, he stopt his works at Bersham, which whether intended or not was a cruel stroke at our business, for engines which we had contracted for could not be executed in due time for want of the castings & the Coalbrookdale Company to whose solicitations for part of our business we had always given a refusal on account of our connexion with him, have not shown much alacrity in executing our orders when necessity obliged us to apply to them ...

Watt's notes look like an internal company memorandum to accompany the list of pirate engines compiled by his son. As such, whether or not John actually saw them is not known, but he will have picked up their tenor and intention from the talk of the time and clearly saw it as a declaration of war. He was not slow to respond. He sent to Boulton and Watt a copy of the advertisement 'I purpose putting into the various Newspapers in different parts of the Kingdom':

STEAM & FIRE ENGINES
The public are hereby informed that Mr. Wilkinson proposes to erect Engines & supply the Materials for Engines upon a new and improved Construction – and the public are also informed that he will indemnify those persons who make use of these Engines from any obstructions that may be thrown in the way by Messrs Boulton & Watt.[39]

The letter in which this intended advertisement was enclosed reflects a keen sense of betrayal and of injustices born by John over the years with great patience and restraint. He talks of Boulton and Watt's recent conduct towards him, effacing the memory of his services to them over the last twenty-two years 'particularly to the former part of it – when … you must be well aware that the assistance I gave you … was the means of carrying you through many of the difficulties which presented themselves in the bringing your Engine Scheme to perfection'.[40] In particular his suggestions for overcoming the practical difficulties involved in the condenser, which in the production models improved the engine but then did not conform to Watt's specification in his patent.

He reminds Watt of the tender ground he has since stood on with regard to this matter and of the dangers of coming to trial over it. This is an implied threat given the then contest with the Hornblowers, and it is in this context that John expressed himself bewildered by Boulton and Watt's conduct towards him 'during the late Contests' with his brother. Before the compulsory settling of his account with the company after the sale of Bersham, he said he had picked up rumours that they had intended to close his works if he did not pay in full, yet 'it was my Intention when that settlement was made to open the Account again provided your conduct towards me afterwards did not [sic] tend to close those wounds in our Friendship which had been opened – This has been very far from the case …'[41]

These exchanges signified the end of any further co-operation between John and Boulton and Watt, whose affairs came increasingly into the hands of their sons. John at this point in his life had no surviving children and within his empire it fell to his nephews, Thomas Jones, William Johnston and Richard Watson, to provide the next generation of service and support. He had clearly discussed this issue in his correspondence with Joseph Priestley in America. Priestley had expressed particular confidence in Richard Watson, whose parents and family the Priestleys knew well, and he assured John that his late sister Mary had always looked upon young Richard as a son.

How confident John himself was of placing his affairs in the hands of these young men is uncertain – the more so since the Wilkinson side of the Priestley correspondence has never been found. Priestley's letters to John show that the two men clearly shared confidences, but John's own thoughts and feelings through this critical period of quarrels and disagreements as he entered old age are missing and this limits a fuller understanding of the events that followed in his last years.

QUEST FOR AN HEIR

After the death in childbirth in 1786 of his daughter Mary (the child of his first wife Ann Maudsley) and since his second wife Mary Lee, who he married when she was 40, had been unable to provide him with children, John's concerns about an heir to succeed him deepened.[1] All his contemporaries – Matthew Boulton, James Watt, James Stockdale, Richard Reynolds and Joseph Priestley – had sons, and for them the new generation was in place to ensure the succession and the continuation of their affairs.

Not so with John. He had no sons. He did have nephews who had become significant figures in his business world and who clearly had expectations for the future. How far such expectations were encouraged by John, perhaps to secure at least some loyalty through his later years of quarrels and litigation, is not clear. He certainly promoted his relations Thomas Jones, William Johnston and Richard Watson into positions of management responsibility in his business world, yet he retained for himself the absolute control of important issues like policy, expansion and capital investment. It was in the nature of the man to retain such control, simply because he believed that he was more skilled, experienced and efficient than they were. It did not necessarily indicate a lack of trust in his nephews, though events ultimately made clear that it had rankled with them.

It was also in the nature of the man to try to solve for himself the problem of having no heir by a carefully constructed process of direct action to procure one, which itself must have contributed to the further insecurity of his nephews. He wanted a son from his own loins and clearly felt that in the late 1790s, as he approached 70, he was still vigorous enough to produce one. But he also needed a woman who would willingly accept his attentions and the clear conditions he would impose, without being difficult and demanding about the consequences.

Most commentators have condemned the process he eventually put in train, by which a mistress bore him three children in his old age, as a betrayal of any Christian morality he claimed for himself and more particularly as a clear betrayal of his wife Mary Lee.[2] But Wilkinson always equivocated in the matter of his Christian beliefs and religious dogma would not have been a serious consideration in his plans. More important to him was his continuing relationship with Mary Lee, and there is good evidence that this was not the betrayal it seemed.

That he cared deeply about her in the 1780s and '90s, when she stayed behind at their northern sanctuary of Castle Head as he went south on business, is clear in many of his letters to James Stockdale who lived at Cark nearby.[3] He refers frequently to letters he has written to Mrs Wilkinson, telling her of his safe arrival at various destinations after days on the road, sometimes asking his friend to deliver an enclosure to her personally that she may have it the sooner, or asking him to confirm that she has recovered from her latest indisposition and that she continues to take her medication. That evidence, extended over the years as they grew older together, is of a close and loving relationship.

Unfortunately, only one of his letters to his wife appears to have survived, but it is particularly illuminating. He first asks her to pass his compliments to Mr Stockdale with some better news of affairs in Cornwall where their investments were not doing well. Then he says:

> … A new mixture of fat and lean, sweet and sour or of good and evil seems to be preferable for us upon the whole, to that choice which if left entirely to ourselves we might make of having only the former without any seasonings of the latter. That you may have and enjoy this in such proportion as to make you happy is the sincere wish of
> Yours affectionately
> John Wilkinson[4]

The letter establishes that his wife was not only a confidant in his business affairs with whom, as with his friend James Stockdale, he could philosophise and discuss matters freely, but also from the tone it is evident that he cares deeply about her happiness.

That caring continued until she died in 1806, still at Castle Head, by which time his mistress had produced the three children, all living with her at his Denbighshire estate at Brymbo. There are no indications that the mistress and his wife ever met, but there is clear information as to how he felt about the loss of Mary Lee. The memorial to her on the south transept wall of Lindale church near to Castle Head, erected soon after she died, has a telling footnote: 'She was humane, liberal and beloved.' That is more than convention required. If she had been simply a wife put aside, the handsome memorial itself, without the footnote, would have been enough to satisfy appearances. And there is more.

Mary Lee left a will, which of itself is significant since it is evidence that she assumed she had property to dispose of on her death, when under the then law any property a woman owned at the time of her marriage, and on her death, belonged to her husband.

Clearly between themselves John and Mary Lee did not accept this process since he had ratified her will in the presence of three witnesses on the day it was made, 16 October 1802. Nonetheless, the law put aside Mary Lee's will on the grounds that she 'had no power to make a will as her estate was the absolute property of her husband John Wilkinson'.[5] Far from accepting this decision John went to law again in 1807, the year before his own death and the year after Mary died, and agreed to the lease and release of all her lands listed in the will to allow it to be executed, because of his 'regard and affection for his late wife and a desire to execute her wishes'.[6]

In the context of this close and caring relationship the two may long have considered and discussed together their lack of an heir, and it is probable that the plan to use another woman to procure one was arrived at jointly. Plenty of wealthy men of the day kept a mistress and maintained them comfortably, sometimes with the knowledge of the wife who accepted the situation provided the mistress did not threaten her status or position. In such cases the mistress was kept for reasons of sexual gratification because the wife could not, or for whatever reason would not, satisfy the husband's physical needs.

It seems to have been different with the Wilkinsons. This one mistress apart, there is no good evidence that John was a philanderer who chased after other women, in spite of his physical vigour which endured into old age and the second-hand stories about his lechery that have circulated both then and since. It is clear, conversely, that he was liked and respected through his middle life and the years of his rise to power by the wives of his friends and business associates for his supportive attitude to women in a society where they had limited status. It is probable, then, that he used this one woman with the agreement of his 'liberal and beloved' wife simply to procure an heir.

The mistress was a maid at Brymbo called Ann Lewis. As such she would see little or nothing of Mary Lee at Castle Head 100 miles to the north, the more so since Mary rarely left her northern domain in these later years, which was probably a strategy of her choosing if she was party to this process. It is likely that the maid was identified as a suitable woman for the purpose by a certain James Adam, also of Brymbo and a man increasingly to be trusted by John who ultimately made him a trustee and executor of his estates under his will. Events proved that to be an unwise decision and a major misjudgement of character; which immediately raises questions as to precisely what relationship existed between Ann Lewis and James Adam at the time and later, the more so since they both lived at Castle Head for most of the twenty years immediately following John's death.

John bought the Brymbo Hall Estate with 500 acres of land in 1792, so it is likely that Ann Lewis became his mistress after that date. In the following nine years to the birth of their first child in 1801 almost nothing is recorded about the relationship. It is an indication of the discretion with which it was conducted, particularly since this was the period of litigation with his brother and his quarrel with Boulton and Watt; a time when he had detractors aplenty who would not have hesitated to abuse his name had they found cause.

There is a dearth of primary information about his relationship with Ann Lewis during the early years she was with him. There are passing references in the later letters of Gilbert Gilpin, whose judgement and commentary has to be suspect following his departure from the Bersham works in 1795 and his disparaging remarks about his former employer elsewhere. One letter in particular, referring to an incident in 1804, is invariably quoted in secondary sources and provides a glimpse of Ann Lewis and of John's responses to her:

> … He has lately been over at B Rowleys for a few days together with his girl. She (poor creature) while there, had nearly died of indigestion from having gorged herself with eating salmons. Old Shylock and her withdrew from the table; and having laid on the bed together for a few hours she returned perfectly recovered. Mr and Mrs Blakeway were there at the same time …[7]

The fact that Mr and Mrs Blakeway were present on this occasion is significant. Edward Blakeway had married the sister of Mary Lee, John's wife, and they and the Wilkinsons had been very close for many years. She was dead by this time and Blakeway had married a much younger woman, but they were clearly still all friends, and the Blakeways were in a position to monitor what was happening between John and Ann Lewis and to report back to Mary Lee at Castle Head as appropriate.[8] The fact that John appears to be comfortable, even uninhibited, by the attendance of close family is also relevant. However, there is no mention of Ann Lewis in the other important sources of the period, either in the letters of Joseph Priestley (who as a divine might not have been sympathetic and therefore might not have been told) nor in the Stockdale family papers.

There is, however, one piece of evidence which suggests that Ann Lewis might not have been the only young woman to have bestowed sexual favours on John in his late years. It is found in John Randall's *The Wilkinsons* in a long verbatim quote from a gossipy letter written by Gilbert Gilpin to William Wilkinson. Randall typically does not give the date of the letter, but he clearly had it in front of him at the time of writing. The present whereabouts of the original are not known and it does not appear in the collection of Gilpin letters held at Shropshire Archives, but in it, in his inimical gossipy style and following on from a string of disparaging remarks about his former employer, Gilpin describes how security was breached at a house where John kept a group of three girls:

> ... Will Rylands and Morris got into his seraglio in the night sometime ago, and the girls (3 in no.) not having full confidence in each other so far as related to keeping the secret, disclosed it, and one of them wrote J.W. respecting it, in consequence of which he wrote Mr. Giles one of his clerks (a seafaring man) to sleep in the house every night, since which Mrs Giles has become jealous of her husband and the ladies of the harem. It is not known how Rylands and Morris will get off when J.W. arrives ...[9]

Gilpin does not say if Ann Lewis was one of the three girls in question and gives no further details, but unless he was making up this salacious story as a gratification to William Wilkinson, whose enmity towards John was by this time well established, it is possible that in his search for a suitable person to bear his heir John might have had experimental relations with other women before he settled on Ann Lewis.

If his relationship with her began soon after his purchase of Brymbo in 1792 there is also the possibility that his getting Ann Lewis with child did not come easily. It may be, of course, that the relationship did not begin until later; their first-born, a daughter, was not born until 27 July 1802 when Wilkinson was 74 years old. He called her Mary Ann, the names of her mother and his beloved Mary Lee. It may be significant that he put his wife's name first.

The second child was also a girl, born three years later on 6 August 1805. How he must have longed for a boy to have called her Johnina. Then at last the following year when he was 78 years old, when he must have begun to despair, on 8 October 1806 a boy was born, and of course he called him John. It is not known whether Mary Lee, by then an old lady in her 80s who died just two months later and who must have been failing at the time, ever saw this baby boy upon whom her husband had set such hopes.

As far as is known all the children were born at Brymbo, but a detailed search has revealed in that area no church records of the births. That is not surprising for two reasons. First, entries in baptism registers of the day recorded children born out of wedlock as the 'bastard child' of the mother, with the father named if known. One can imagine such implied condemnation being anathema to John. Second, his known sympathy towards dissenters and his open criticism of the crippling power of the Church on those who did not conform to its dogma might have precluded baptism in the Church of England anyway.

In these last years, in his late 60s and 70s, John still continued to be actively involved in the management of his business world. He spent more time at Castle Head during this period and he allowed his nephews, particularly Thomas Jones, more scope and gave them more support, but he continued his drive to extend his empire and added to his concerns whenever he saw an opportunity. Perhaps his acceptance in 1799 of the only public office he ever seems to have held, that of High Sheriff of Denbighshire, was because he felt it would help retain his status and reputation and avoid a fall in his fortunes in consequence of his relations with Ann Lewis. He may have dreaded the decline which would have led to a lesser inheritance for his heir. The relinquishing of his Broseley home, The Lawns, leased from the Forester estate, also occurred at this time.

His ironworks continued to innovate, however. By 1800 at Bradley 'great additions and improvements have been made recently in separating the dross from the ore by using huge concave rollers instead of hammers'.[10] The steam-driven forge hammers there were themselves the innovation of 1783, when John employed Watt's new rotative engine to raise forge hammers for the first time.[11] The huge concave rollers were something new. Moreover, at Hadley soon after the turn of the century in 1804 he built two new furnaces on his estate there, purchased in 1791 but undeveloped until that date.[12]

Perhaps the best indication of his continuing determination to keep his name in the forefront of the iron-making industry, however, was his involvement in April of 1801 in plans to build a new Iron Bridge, this one in London, over the Thames. The contact came through Thomas Telford, who for many years had been an admirer of John's drive and the quality and skill of his iron-making.

The two men had met in the early days of Telford's appointment as county surveyor for Shropshire in the late 1780s; Telford then a young man in his late 20s at the beginning of his career, John already a wealthy and established ironmaster. Telford was mainly responsible for public buildings and important houses, but soon moved into bridges and is likely to have come to John's notice with his proposal to build the new Buildwas bridge across the Severn, just above Coalbrookdale, in a single arch of iron. It is probable that John recognised the energy and confidence of this young man as a kindred spirit whose roots and aspirations were similar to his own. Certainly, he was supporting Telford in 1793 for the post of engineer to the Ellesmere Canal against a second candidate put up by a group of the commissioners. As Telford wrote to his former school friend Andrew Little in Scotland:

... I deferred answering ... untill I could let you know the determination of the general meeting of the Ellesmere Canal Navigation, which was held last Wednesday. They have confirmed my appointment as general Agent ... they

Thomas Telford, whose talent as a civil engineer was recognised and supported by John Wilkinson.

had endeavoured to raise a party at the general meeting, but we were too powerful for the opposition ... I had the decided support of the great John Wilkinson, king of the Iron Masters, who is in himself an host. I travelled in his Carriage to the Meeting and found him much disposed to be friendly ...[13]

The contact between the two men continued because John tendered for the ironwork required for the water bed and for the arches linking the huge stone pillars in the Pontcysyllte Aqueduct – another innovative Telford undertaking carrying part of the Ellesmere Canal 127ft above the River Dee. John's tender was too high for the commissioners, but Telford returned to him again a few years later in connection with the London Bridge project.

Telford with his then partner, Douglas, had been commissioned by a committee of the port of London authority to examine the options for a new bridge over the Thames as part of a proposed extension of the port facilities, and had proposed a bridge in iron with a single span of 600ft. A list of mathematical and technical questions about the structure of such a bridge had been prepared by Telford, who must have indicated to the committee that John was the man to advise them. John's letters on the subject to Thomas Telford and Samuel Gunnel, secretary to the committee, and his technical advice in response to Telford's questions are preserved in the archive of the Institute of Civil Engineers in London.[14]

Several interesting points emerge from these documents. First, either Telford or the committee, or both, had decided to seek specialist advice on the technical and financial aspects of construction and on the nature of the iron to be used from John, rather than from the Darbys of Coalbrookdale who twenty years previously had built the first Iron Bridge over the Severn to such wide acclaim. The committee did, however, consult William Reynolds for his iron metallurgy knowledge.

It is clear that in addition to John's letters to the committee a separate correspondence was carried on with Telford on more detailed technical points. The correspondence also revealed that some persuasion might have been necessary to encourage the committee to proceed, with England at war with France and Napoleon gaining victory after victory in Europe. The bridge was in fact never built, but the whole episode is a measure of John's continuing importance and evidence that his reputation as an ironmaster was still intact, whatever the rumours circulating about his private life.[15]

There are no signs here of physical or mental decline. His use of language in these exchanges is masterly: simple, concise and, even in discussing uncertainties, his meaning and intention absolutely clear. The handwriting, too, remains strong and rounded; easily and clearly readable. It would be helpful, of course, to see the portrait of him known to have been painted alongside William Reynolds and Samuel Crawshay Junior in 1796 by a Mr Wilson of Birmingham. In a letter to Matthew Boulton he says 'the likenesses are deemed good, mine in particular is said to be a very strong one'.[16] Unfortunately, the whereabouts of that portrait are unknown. The well-known Abbott portrait of him is from an earlier period.[17]

Shortly after the birth of his long-awaited son, John applied for a coat of arms; more evidence of a wish to establish conventional status in the context of his children's future. The arms were eventually granted to the three children in 1808 after they had assumed, by royal licence, the name of Wilkinson. The immediate concern, however, was the will.

Following the birth of his son John lost no time in publishing it, on 29 November 1806, six weeks after the birth and just three weeks before his wife Mary died.[18] She could have been party to it; their last important shared act together. The terms of the will outline the way they had been thinking in the quest for an heir during their last years.

An interesting point to emerge in a codicil following Mary's death is that John clearly wished Ann Lewis to live as his widow with his children at Castle Head following his own death. It is an indication of his continuing attitude to this, his northern sanctuary, to which place his letters show he always yearned to return for the peace it afforded him from that frantic, noisy world where he made his money. He clearly wanted his children to love this place, perhaps Ann Lewis, too. But living on at Castle Head with another man if she remarried after his death was specifically excluded. She retained the benefit and an annuity only so long as she remained unmarried. If that happened it reverted to the children.

Whether or not she came to live at Castle Head in those nineteen months before he died is uncertain. If he wished to have his children there in the closing period of his life it is probable she did come, given a decent lapse of time following Mary's death. The summer of 1807 seems a likely time. Since John spent most of his last year at Castle Head the wording of the codicil seems to confirm she, too, was there when it refers to 'the children which he might have by the aforesaid Ann Lewis and living at his decease or born within six months after'.

Good provision was made for Ann Lewis in the will and a fall-back position was established, whereby everything went to his nephew Thomas Jones (provided he took the name of Wilkinson) if none of his children survived. Its main thrust, however, was to ensure the continuity of his iron-making empire for the ultimate benefit of his children by Ann Lewis. To achieve this, a trust was established to manage the whole of his estate, with trustees given wide freedom to buy and sell as they thought best for the next twenty-one years (amended by a codicil from the thirty-one years he had first postulated). Clearly, John believed his many works, most notably the large and thriving ironworks of Bradley and Brymbo which were specifically emphasised, would survive and develop during this period and become an increasing asset. Yet given the broad terms under which the trustees were permitted to operate, everything depended on their management and on their integrity.

He appointed as trustees and executors Ann Lewis, James Adam, his old works manager Cornelius Reynolds and Samuel Faraday. There were changes almost immediately. In a second codicil to the will John revoked the appointment of Cornelius Reynolds and appointed William Smith in his place, following which he made a list of reasons for appointing each trustee and what their probable duties would be. Samuel Faraday was then declared bankrupt, withdrawn as trustee and escaped from the jurisdiction of the court to France. It left the Wilkinson empire increasingly in the hands of Ann Lewis and James Adam. The wasting process was about to begin.

POSTHUMOUS RUMBLINGS

John died on a visit to Bradley on 14 July 1808. There is no record of a declining illness. *Aris's Birmingham Gazette* on 18 July carried the following notice of his death:

> Thursday, at his works at Bradley, in the County of Staffordshire, at the advanced age of 80 years, John Wilkinson, Esq. Few men are more entitled to the praise and gratitude of their country, for unwearied and successful exertions in raising that important branch of our national production, the iron trade, to a height unknown, until that period which constituted the zenith of his powers. Frugal, though not parsimonious, he acquired an immense fortune, presenting to society the satisfactory testimony that, in this free and happy country, industry and prosperity go hand in hand. The loss of such a man, considered in his multifarious connections with the manufacturing class of society, must be great indeed; but the calamity will be in some measure palliated, as a very efficient Trust has been appointed to carry on his vast and extensive concerns.[1]

In the event the very efficient trust proved unfit to carry on his vast and extensive concerns, in spite of the resounding words he wrote in his instructions to the trustees: 'I leave my different works as children in trust ... that a great example may be given of the importance to the world and benefit to the industrious workman arising from infant works being protected until their arrival at a proper maturity.' But there were many complementary and related issues. Further instructions to his executors laid down what should happen to his body:

> ... It is my particular request and direction that wherever I die my body may be interred as privately as possible without parade or pomp, either in my garden at Castle Head, within a place I have there prepared for that purpose, or within a building called the chapel at Brymbo, or in my garden at Bradley, in such manner as is directed in this book ... and to the nearest of the said places I shall happen to die ...[2]

Since he died at Bradley in Staffordshire close to the site of his great ironworks it is interesting to ponder the reason why his executors, in view of the above instruction, chose to undertake the difficult task of transporting the body, and presumably the huge cast-iron grave monument too, 100 miles or more over villainous roads to Castle Head. It may have been simply because a place was there prepared for that purpose, which might not have been so at Bradley or Brymbo and the executors thereby thought to save themselves some trouble; or it might have been because Castle Head, far from the crowded industrial world of the Midlands, afforded the opportunity to inter the body 'as privately as possible without parade or pomp'.

Whatever the reason, to Castle Head the body came: on a gun carriage, appropriately enough, and in a lead coffin encased in a wood one (according to Stockdale), behind two horses in the charge of two men. The story of the surprising journey is nowhere told in primary documents of the time and has only passing references in Stockdale's annals, but the folk memory of it in the Castle Head area is still strong and still recounted.

It tells us that by the time the carriage reached Lancaster it had been on the road for four days. It would necessarily have been a slow and solemn journey. At Lancaster the men decided that instead of taking the long route northward round the bay and the fringing mosses they would head out across the sands direct for Castle Head; the route John himself always took when he came here. Unfortunately they did not check the tide times before they left and a mile or two short of the Castle Head shore found the incoming tide running quickly up the channel alongside the sandbank they were following. Realising they were in danger of drowning they quickly loosed the two horses from the traces, climbed on their backs and galloped to safety ahead of the rising tide, leaving the gun carriage with the coffin strapped to it standing on the sandbank. On reaching the shore they turned and watched in horror as the sea rose around the carriage, lifted it and tipped it sideways so that the coffin broke away and floated free for a moment before it sank below the waves and disappeared. It was late evening and almost dark when they reached the mourners waiting at Castle Head and told their dismal tale.

Early next morning an old gardener from the house went down to the shore to check his fish nets before the next tide and found the coffin sitting on the rocks and still intact. He hurried back to Castle Head shouting that he'd found the master. Quickly a horse and wagon were brought and a procession of excited people followed behind to the shore and helped bring the coffin back. It was carried quietly to a ride in the wood beside the intended grave and placed on the ground there until the burial proper could be arranged.

The story usually diverges at this point to recall the words of an old gypsy woman who, years before, had stopped John's carriage and almost thrown it over. It seems the coachman had seen what he thought was a bundle of rags in the road ahead, which had risen up suddenly in front of the horses to make them rear and shy. There was commotion inside the carriage, with John's head shouting out of one side window and his friend, said to be Samuel More, looking out of the other. Regaining control of the horses the coachman tried to whip the old woman out of the way but she evaded him and pointing straight at John shrieked 'that man ... that man' and turning shuffled off down the road cackling to herself. John's companion who had alighted from the carriage heard this, and followed after her asking what she knew about that man. The coachman overheard the reply. 'That man,' she said, 'that man, must be buried four times before his spirit will rest', and turning down a path into the wood alongside the road she disappeared. No one thought any more about the incident at the time, but the coachman recalled it the night after they thought John had been buried by the sea.

It is worth examining here Samuel More's description of the burial place at Castle Head that John had prepared for himself. As early as 1783 he had decided he would be buried in an iron coffin and began to prepare the place for it on Castle Head Hill. More describes it as follows:

> The Place designed to Receive the Remains of Mr. Wilkinson (when he shall be called from doing Good on this Earth to receive that Reward which such as he is have [sic] every Reason to expect and no Doubt of Obtaining) is now completed. Six Cases of Cast Iron are placed in a Cavity cut in the Rock to receive them and in them are to be deposited Mr W. and Some select Friends in wooden Coffins, the Ends of the Iron Cases being contrived to take off and may be screwed on again with such Inscriptions upon them as may be thought proper to commemore [sic] the Person enclosed. It was at first intended that this Place of Rest should have been in Manner of a Cave but that Idea is now changed and Trees and Shrubs are Planted which in Time will form a Grove ...[3]

The first point to make is that in 1783 John was at the height of his power and influence and in his own mind could reasonably expect that 'select Friends', in perhaps the dissenting if not in the Quaker community, would wish to be buried here in his little piece of paradise. Samuel More neither indicates who those friends were nor if they had been invited.

Second, it is clear that the bodies were intended to be brought here in wooden coffins, which the iron coffins were large enough to accept, with an open end that could be screwed down later. This pre-supposes an iron cask perhaps a couple of metres long by something like a metre across; but were they cylindrical, or rectangular in the traditional shape of coffins? John's lateral thinking might well have evolved a novel use for his new cylinder-boring technology.

The final answer to that question is unlikely to be settled until the Carlisle diocesan authorities grant permission to open the Wilkinson vault beneath the floor of Lindale church. So far this has been refused, following an initiative by the parish council to locate the vault precisely using sophisticated metal detecting equipment. A very strong signal has been obtained over an area which suggests there may be two iron coffins there, which of course raises a further question. Was his wife, the 'humane, liberal and beloved' Mary, buried only eighteen months before him, also entombed in iron?

But to return to Samuel More's 1783 description of the grave site on Castle Head Hill: there is a third point. He tells us that the hard limestone rock had to be excavated to receive the coffins, not in the manner of a cave as was at first intended, but to form a grove, presumably with individual excavations made down into the rock from the surface and separated by shrubs and trees. It is important to note here that the site of the grave, as marked by the iron obelisk on a painting made shortly after the burial, places it on the line of a major geological fault passing through Castle Head Hill, where the rock would lend itself to easier excavation from above rather than to a cave structure where the roof would have been unstable.[4] John's contacts with the Lunar Society and their geology researchers, particularly in the Buxton area, would mean that he came to realise this.

The folk memory has it that when the Castle Head mourners assembled at the grave site for the burial proper, the lead coffin inside the wood one which held John's body was too big to fit into the iron coffin awaiting it. A new, larger iron coffin was sent for from his works and since this was some time in arriving the body in its enclosing containers of lead and wood was left sitting in the ride under the trees beside the grave site.

Days later when the new iron coffin arrived there were further difficulties. This larger container was now too tall to sit in the excavation in the rock that had long been prepared as the grave, so again the body was left in the woodland ride and the rock-cut grave was enlarged and deepened before the burial could be held and the grave sealed. The huge cast-iron obelisk bearing his epitaph was then erected above it.

There he lay for twenty years, as conspiracies were planned by executors and confusion grew among beneficiaries. The writer Graham Sutton in his novel *North Star* gathered together the folk memories in the Castle Head and Lindale area associated with John and incorporated them in an imaginary account of the burial.[5]

In view of Samuel More's words in parenthesis in describing the grave site, it is worth mentioning here the nature of John's grave memorial and the wording of his epitaph on it. More himself seems to have had a strong and conventional religious faith, attending church on Sundays wherever he happened to be on his journeys through England, but there is no evidence that John ever accompanied him or worshipped in church himself. More nonetheless acknowledged John as a good man whom he believed would have his reward in heaven in due time.

The two men and their respective wives, both spinsters of 40 when they were married and neither able to have children, were the closest of friends for forty years until More's death in 1799. They were frequently together and the Mores were guests at Castle Head almost every year from the time it was built. They shared many confidences and were involved in the 'Philosophick' issues of the day, and it is inconceivable that in this long association the men, and their wives since both were known to be strong personalities in their own right, did not talk together about religion and affairs of the human spirit.

In that context it is likely that More had some input into the wording of the epitaph and the nature of the monument Wilkinson planned for himself. The memorial erected over his grave, later described somewhat disparagingly as 'a pyramidical mausoleum consisting of some 20 tons of iron', was designed to carry plaques containing his epitaph and a relief of his head. The words he wanted ran as follows:

Delivered from Persecution of Malice and Envy, here rests John Wilkinson, Ironmaster, In certain hope of a better state and Heavenly Mansion as promulgated by Jesus Christ in whose gospel he was a firm believer. His Life was spent in action for the benefit of man, and he trusts in some degree to the Glory of God, as his different works that remain in various parts of the Kingdom are testimonies, of increasing labour, until death released him date/ age etc ...

What he got was subtly different. Since More was dead by then; the Priestleys dead; Mary Lee gone and unable to speak for him; and his humanity and certainly his religious belief scorned by his detractors because of his affair with Ann Lewis, it can only be that his executors and trustees, with Ann Lewis and James Adam chief among them, decided that a modification of the wording was necessary for appearances' sake. Since John's epitaph plaque would certainly be ready waiting at the time of his death this must also mean they went to the trouble of casting another. John's own words on the original were in all likelihood physically, but certainly metaphorically, melted down to create the new plaque, which said:

> John Wilkinson, iron master, who died 14th July 1808 aged 80 years. His different works in various parts of the kingdom are lasting testimonies of his increasing labours. His life was spent in action for the benefit of man, and, as he presumed humbly to hope, to the Glory of God …

The altered words remain on the plaque to this day. It was the first of many manipulations of his wishes by the executors.

There is no entry in the Cartmel register of burials for John's death and interment, but an entry in the Lindale register reads: 'July 26th. John Wilkinson of Castlehead Esquire aged 80 years. N.B. After the Funeral Service was read his remains was removed by order of his Executors and deposited in the garden at Castlehead.'

The question that emerges from this, of course, is did John seek in the last stages of his life, in spite of his dissenting views about conventional religion and perhaps remembering some earlier persuasive arguments from Samuel More, to have the burial service read over his remains? It is more likely to have been arranged by the executors, again seeking to minimise outrage and avoid any future difficulties over the unconventional procedures. The absence of a proper burial record, too, would have complicated the proving of the will so there could have been no more to that than the need to satisfy a legal nicety.

A further piece of evidence does suggest, however, that Ann Lewis did not feel herself bound by John's wishes, if not in this instance his legal instructions, in her decision to have at least two of their children christened in church almost immediately after his death. Perhaps she made this move against the dissenting principles of the father in almost indecent haste to achieve a semblance of respectability as the new mistress of Castle Head. It might indicate, however, that she was not in sympathy with him at the end; that she resented something at the time of his death.

There are two consecutive entries in the Cartmel register of christenings for 29 September 1808, indicating the christenings took place at the same time.[6] They read: 'Johnina daughter of John Wilkinson Esquire and Ann Lewis, Spinster born 6th August 1805, christened 29th September 1808'; 'John son of John Wilkinson Esquire and Ann Lewis, Spinster, Castlehead. Born 8th October 1806, christened 29th September 1808.'

There is nothing in the register about illegitimacy, which might say something about Ann Lewis' powers of persuasion. Perhaps to record her as spinster was enough to satisfy Church conventions. In the second entry she becomes 'Spinster, Castlehead', which suggests she was certainly living there by then and perhaps had been for some time. There is no mention here of the firstborn child, Mary Ann, who would have been 6 at this time and might have been christened elsewhere.

The children grew up at Castle Head and lived there with their mother as they grew to adulthood, perhaps as their father intended. What kind of a house did Ann Lewis keep? Did the children have a governess, a tutor, a social life? Under the amended will following Mary's death, Ann Lewis held Castle Head, its furniture and plate and an annuity of £500 so long as she remained unmarried. There was also an allowance which she controlled of £200 a year for the welfare and education of each child, and a further annuity to herself of £200 for supervising the children's affairs, so she was a relatively wealthy woman.

James Adam was soon in a position to become the controlling trustee of the estate, which allowed him great freedom to buy and sell. The more so since Ann Lewis herself seems to have been the only other active trustee from an early date, so the two of them might have been able to divert further resources to their own affairs. They had, of course, known each other throughout the time Ann Lewis was with John. Precisely what their relationship was before and after his death, and throughout the time the children were growing up at Castle Head, invites speculation.

As they grew older the children would learn they were infant wards of the court of chancery and the ultimate beneficiaries under the will of their father, whose large cast-iron monument and mausoleum stood facing them throughout their childhood years across the lawn and the south-facing windows of the house. It might have cast a long shadow across them, but their comfortable lives and the apparent security of their future would have done much to disperse it. That future and the inheritance it promised were soon threatened, however, by the activities of the said James Adam and by Thomas Jones, John's nephew, the son of his sister Sarah.

Using the freedoms accorded to the trustees in the will, James Adam, by now the controlling trustee, began to buy land all round the Castle Head estate in the Winster Valley.[7] He bought the land in his own name but used the trust as security for his mortgages. Whether because of other expenses or extravagances he was involved in at this time, or simply because of poor management and accounting, he fell heavily into debt and clearly needed a substantial sum of money. It is likely that at this time he learned through his involvement in trust affairs that money could be raised from the government under an Act of George III. He subsequently applied and obtained from the Crown Commissioners a loan of £35,000 in exchequer bills. For security he provided a bond of £70,000 on the Wilkinson trust. It was to be his undoing.

What he did with the £35,000 is not clear, but it certainly did not solve his problems. He fell further into debt and his final downfall came when the government called in their bond. In 1822 he petitioned the high court of chancery in a submission which stated that he had contracted large debts in conducting the trust's affairs and that by selling some of the trust estate a sum should be raised to pay them. The petition had not been heard by the time of his death the following year.

In his will he repeated this argument and again tried to pass his debts on to the trust by stating that, although he had bought property in his own name, he had intended that it should all ultimately be conveyed to the trust estate at the prices he had paid for it initially. When his petition was eventually heard the court judgement held that in view of his substantial debt to the Crown, which was now called in, it would not be to the advantage of the trust to adopt the lands bought in this manner. As a king's debtor on his death, all his estates were confiscated by the Crown.

It is significant that in the year following James Adam's death Ann Lewis married a certain Thomas Milsom. Perhaps she saw the end coming, but it would not be done lightly for at a stroke she lost her right to live at Castle Head and her £500 annuity; Thomas Milsom was perhaps a man of some means. She'd had, of course, the opportunity to put something by for her later life through the thirty years of her contact with John and the Wilkinson trust! The Castle Head estate, by this marriage, came directly into the control of the Wilkinson trust and Ann Lewis's name at this point disappears from the record. Whether or not she continued to act as a trustee is unclear.

Meantime, Thomas Jones, as a potential beneficiary named in John's will to whom the estate would pass if all the children of Ann Lewis died (provided he took the name of Wilkinson), had started litigation of his own. It is worth pointing out at this stage that Thomas Jones' costs and expenses as a potential beneficiary, and certainly the trust's costs in opposing him, could be charged to the trust estate.

As a principal manager of his uncle's iron-making affairs, particularly of his estates at Brymbo, and as a close relative who had had the confidence of his uncle for years, Thomas Jones contested the will as soon as it was proved on the grounds that the children were unnamed in it, and in any case were born out of wedlock. It took four years for his submission to come to trial in the court of chancery on 13 March 1813. Secondary sources have reported that when the judge hearing the evidence, Lord Eldon on that occasion, asked him what provision he intended to make for the three children if his petition were successful, Thomas Jones replied, 'None.'[8] It seems an unlikely scenario. Thomas Jones was an astute and educated man, and unless some great rage had continued with him through the five years since his uncle's death, and betrayal if he so perceived it, he would have been more mindful of the possible outcomes of this response. Whatever may have happened in that court session, a decree was issued dismissing the petition of Thomas Jones Wilkinson.

Without delay Thomas Jones then sought a decree against James Adam on the grounds of his mismanagement of the trust estates; that he be restrained from using trust money for his own purposes. Thomas Jones obviously knew of James Adam's Winster Valley land transactions around Castle Head, and clearly believed it was his own rightful inheritance that was being wasted. Why the two did not come to some accommodation at this point, if only on the merely practical issue of conserving the estate in which they each had interest, might be a measure of the continuing anger of Thomas Jones and his conviction that his claim would be established in the end. It is an interpretation supported by his responses when this decree eventually came to trial six years later in 1819, just before James Adam's debts caught up with him. The Judge found in favour of James Adam whose 'faithful attention to the will of his late master' was extolled, and the second petition of Thomas Jones Wilkinson was thus dismissed.

Jones immediately put in process a final appeal to the House of Lords to reverse this judgement. It took another four years for it to be heard, in which time certain restrictions were in fact imposed on James Adam with enquiry to be made as to his mode of keeping accounts.[9] The legal costs and other expenses to this point must have been colossal.

Although James Adam was known to be in serious financial trouble by the trial date, the unanimous opinion of the judges affirmed the decree in favour of James Adam and others, in other words the Wilkinson trust, but not before they had called counsel for both sides together in one of those secret pre-judgement conferences that are never part of the record. Were they concerned about the continuing costs exceeding the value of the estate? The entry in the House of Lords journal on 11 June 1823 indicates that evidence had been heard from Ann Lewis and all the other trustees named at one time or another in the will, including William Vaughan, Samuel Faraday and William Smith, and from James Adam's son, Samuel Smith Adam, who is recorded as the guardian of the three illegitimate children.[10]

Throughout this period we know little about the comings and goings at Castle Head, where the children would hear and understand more of events and come to form their own judgements as time passed. It would help our modern-day understanding if we knew who the visitors were and what they thought. Inevitably, there would be some entertaining to do in connection with the administration of the trust, which would certainly bring people of the wealthy manufacturing class, former contacts and friends of John, perhaps even the gentry, to Castle Head. Friendships will have evolved between their children and Ann Lewis's children, enduring perhaps into adulthood.

One such set of visitors, the Leghs of Lyme, an ancient landed family who had held Lyme Park for 600 years, was associated with the Wilkinson trust business operations at St Helens near Warrington. There might have been more than merely a business connection to Castle Head and the Wilkinson trust, for the Lyme Park estate in the 1790s was held by a certain Thomas Peter Legh, another dissenter and free thinker. About the time John was associating with Ann Lewis this nobleman fathered seven children to maids on his estate, four of them in one year. He never married but acknowledged all the children and had them brought up and educated at Lyme Park. It may be no more than coincidence that he and John were concurrent in these affairs and had over-lapping business interests; but it is curious that one of these children was William Legh, who must have met Mary Ann Wilkinson when she was a teenager, and married her in 1821 at Cartmel near Castle Head when she was still only 19.[11]

He had to obtain the consent of the Lord High Chancellor to do it, which perhaps would not be too difficult for a family of his standing. There was, too, the question of Mary Ann's expectations, her dowry by that time less certain as heiress under the Wilkinson trust, which had become the focus of such increasing litigation. William, however, managed to formally execute articles that contained a power for him to draw on the trust for an annuity of £500 a year following the marriage, or once only for a sum of £10,000 in the period until the trust was finally determined in favour of the three children. He chose the second option, aware no doubt of the litigation involving the Wilkinson trust current at the time between James Adam and Thomas Jones and of its uncertain outcome. Part of this settlement was the Castle Head estate.[12]

It is not clear if Ann Lewis was still living at Castle Head in May 1821 at the time of the marriage of William Legh and her daughter Mary Ann, but since the newly married couple lived there for at least five years thereafter it seems unlikely. Mary Ann, as a young bride, would not wish to be under the close surveillance of the former mistress of Castle Head, who must have been a dominant mother.

In the next five years two sons were born there to William and Mary Ann, both dying in infancy.[13] A close relative of William Legh, Reverend Peter Legh, then of Newton near Lyme Park, came to stay at Castle Head to christen the firstborn and, ultimately, to take the burial service when his death seemed certain. Memorials to the two dead infants can still be seen on the south wall of Lindale church.

The brief tenure of William and Mary Ann as master and mistress of the place – which had such strong associations with her father; where she met and fell in love and had happy memories – was clearly a sad one, and the indications are it was more than they could bear. John perhaps turned in his grave at the loss of their two sons followed soon after by their decision to leave the place. It is significant in view of the developments within the Wilkinson trust at this time, following the death of James Adam and the marriage of Ann Lewis, that William and Mary Ann moved to Brymbo Hall rather than to Hordle in Hampshire where William held a small estate of the Leghs.

Brymbo, in any case, had important associations for Mary Ann going back to her infancy, and was within easy reach of William's family seat at Lyme Park where they visited regularly. 'Aunt William', as Mary Ann became affectionately known within the Lyme Park circle, became a famous lady archer. In the following years up to the early death of William in 1834 Mary Ann bore three more sons and two daughters, all of whom survived. She died herself, however, only four years after William aged 36, and was destined never to know her grandson who became a great statesman and the first of a line of Lord Newtons of Lyme.

Her short life is full of resonances and echoes of her father: a halcyon childhood at Brymbo and on Castle Head Hill; the shadow of the free thinker, that led to her own, and to her husband's, illegitimacy; the marriage into the landed gentry; the loss of a beloved spouse; the early death of children. Lyme Park is an educational and visitor centre today, which has been interested to publish the Wilkinson links with the ancient family. It is unfortunate, however, that a late Victorian matriarch decided that these links did the family no honour, and with great determination and what can only be described as Wilkinson thoroughness rooted out and destroyed most, but not quite all, of the documentary evidence that existed in the family muniments.

The emerging personality of Mary Ann's sister, Johnina, is less easy to follow. Obvious disappointment that she was a girl might have thrown a shadow across her infancy, the more so since the boy so longed for and, therefore, likely to be shown preference was born the following year and they were reared together as siblings with the older Mary Ann. Who of the three was the leader and initiator during their childhood and teenage years? Later handwriting evidence in the Cartmel registers suggests it was Johnina, whose signature, big and rounded and flourishing, seems to dominate Mary Ann's thin, small, scratchy, uncertain hand.

Johnina was 16 when her sister married, and was still at Castle Head when Mary Ann returned from honeymoon in the summer of 1821. Were they then still young enough and close enough as sisters to share confidences? Sometime towards the end of the Legh's five-year tenure of Castle Head, Johnina was courted by a certain Alexander Murray, a well-to-do Scotsman whose father,

William Murray, lived at Palmaise in the parish of St Ninians, now a suburb on the south side of Stirling. She was married from Castle Head at Cartmel by banns on 12 April 1826.[14] Her brother-in-law, William Legh, was a witness to the marriage and in all likelihood gave her away. He had already, through his family contacts, made the same kind of arrangements with the Lord High Chancellor for Johnina's dowry as he had previously made for himself on his own marriage to Mary Ann: an annuity of £500 or a once-only option to withdraw £10,000. This suggests a continuing closeness between the sisters at the time.[15] Mary Ann herself might not have been present at the marriage of her sister. Just a month earlier she had buried her second baby son.

In the two years between their marriage and the sale of Castle Head out of the Wilkinson trust in 1828, and perhaps as a consequence of an understanding or agreement between the sisters, Johnina and Alexander lived there, perhaps returning to the place from honeymoon after the Leghs had left. After the sale they, too, moved to Brymbo. Theirs, however, was another brief interlude of happiness. Alexander died at Brymbo Hall on 5 June 1835 and was buried at Wrexham where there is a memorial to him on the inside wall of the parish church, not far from the one erected there by John to his first wife Ann Maudsley, eighty years before. With the dispersal of the Wilkinson trust estates, Johnina disappears into the mists.

The third of the children, John Wilkinson Jnr, has been the most difficult of all to follow and the documentary evidence is scarce and patchy. Following his christening it is clear that he was not brought up at Castle Head as a dissenter because he went up to Christ's College, Cambridge, which would not have been an option had that been so. On reaching his maturity in 1827 he obtained an advance of £1,000 from his friend and attorney, James Kyrke of Wrexham, for an intended tour of the continent of Europe – the Grand Tour no doubt, following his Cambridge education. The money was to be repaid from his trust annuity of £200 on the security of his third share in the estate.

Some undocumented research, which has yet proved to be reliable in many other aspects of the Wilkinson story, indicates that in 1829 Thomas Jones Wilkinson, who had bankrupted himself contesting his uncle's will, leased the Brymbo ironworks after the sale of the estate and with the help of none other than John Wilkinson Jnr ran it successfully until 1837.[16] In that year John Jnr is said to have emigrated to America, where he married and had seven children, returning to Brymbo only briefly in his middle age to visit the works and meet the old people who remembered both him and his father.

As early as the summer of 1825 the three children, the co-heirs, had clearly seen that the estates in the Wilkinson trust were wasting, their inheritance diminishing, and that claims against the trust, particularly those presented under the will of

James Adam, continued to be threatened.[17] It is not surprising, therefore, that through the family friend and attorney, the same James Kyrke of Wrexham, they petitioned the court of chancery to have claims against the trust assessed, and where established liquidated by a sale or mortgage of part of the estate.

In the master's report of 1828 it was stated there were justifiable claims against the trust of upwards of £178,000 (in then values) largely contracted by James Adam as trustee, as well as other claims pending and so far not examined. It may be that with the additional cost of litigation the total of costs would be more than the estate could bear. However, in view of the petition of the co-heirs, sale of the trust estate was to begin as a matter of urgency by either public auction or private contract. An order was made to this effect, and James Kyrke was named by the court as the sole trustee with authority to proceed.

A catalogue of all estates and properties was drawn up, listing messuages, cottages, water corn mills, iron furnaces, orchards, meadow, pasture, woods, heath, water, fisheries and other rights; noting the parishes within which each property stood and the names of the tenants to whom properties had been leased. It is a catalogue and map of the Wilkinson empire taken into the Wilkinson trust, though perhaps modified by James Adam.[18]

It is not known which of the interested parties started at this point the move to exhume the body of John Wilkinson to have it re-buried elsewhere. Stockdale, writing in 1870, collected information on the process, very little of it documented, and he clearly used the folk memory of the event which is still around today.[19] It seems that the agents appointed to conduct the sale of the Castle Head estate suggested to the co-heirs that the dominating mausoleum of their father within sight of the drawing room windows would, as Stockdale put it, 'injuriously affect the sale'. Eventually the curate at St Paul's church, Lindale, the Reverend Anthony Barrow, was consulted and proved helpful.

An unusually long entry in his handwriting in the church register is explicit and effectively answers those who have long maintained that the priest would not accept John's body for re-burial within the church, and that he was re-interred in an unmarked grave in a field over the graveyard wall:

1828. Aug 16th John Wilkinson of Castle Head died at Bradley July 14th, 1808 aged 80 years. It appears by the Register of Burials that on the 26th of the same month the burial service of the church was read over his remains at this chapel; after which they were by order of his executors removed and deposited in the garden at Castle Head. Now this is to certify that the said Remains of the said John Wilkinson were again removed and on this 16th of August 1828 brought back again and deposited in a vault in the said chapel. Witness: Anthony Barrow   Minister.[20]

Since a note of this register entry can still be found among the Stockdale family papers, it is reasonable to assume that James Stockdale had it carefully copied out for him by a certain J.H. Ransome at the time his annals were being compiled. There is another verification of the Lindale re-burial of Wilkinson among the same papers, quoting the evidence of a Mr Thomas Remington of Lindale who was 88 years old in 1870 and, therefore, presumably witnessed the reburial as an adult aged 46 at the time.[21] This reference mentions particularly the iron coffin and is very specific as to where it was re-buried in the church: 'in the family vault inside of Lindale Chapel at the North End close to the Communion Table.' Stockdale must have put out a local plea for information on the event.

Anthony Barrow's entry in the register makes clear that the Castle Head tomb was opened and the body removed less than two weeks ahead of the date already set for the auction, leaving little time for any tidying up of the site. The folk memory is that the iron monument which had stood above the grave was left lying in the brambles under the trees on the hill for years and the highly romanticised illustrative poster of Castle Head and its 'gigantic mountain' commissioned by the agents to go with their sale particulars does not show it.[22]

James Kyrke had lost no time in organising the sale of Castle Head by auction. Indeed he was ahead of events, for he had first advertised it for sale in the *Westmoreland Gazette and Kendal Advertiser* on 12 July 1828, a month before the date of the order authorising him to do so. Castle Head house and lands were auctioned at the Crown Hotel, Grange-over-Sands on 28 August, along with Wilson House farm, the home farm and a small estate in its own right comprising messuages, cottages, water mill and 800 acres of land. Low Meathop farm and Holme Island, as part of the James Adam purchases, were listed for the following day. The Castle Head furniture and contents were auctioned on 13 September 1828.

The Castle Head property had been valued at £20,000 but was expected to sell for less and might not, in fact, have sold at auction. It was, however, purchased by a Liverpool attorney called Robert Wright of Stand House, Wavertree, Liverpool. For among his papers was found a receipt for the property dated 20 August 1829 for £6,150, paid to the account of 'J S Harvey, Accountant General for the High Court of Chancery'. That amount from this first of the estate sales under the Wilkinson trust will have provided scant comfort to the co-heirs, with that huge debt to cover before they received any further benefit from the disposal of their father's former empire.

On 23 and 24 April 1829, a little over six months later at the Wynnstay Arms in Wrexham, the Brymbo Hall estate, comprising the mansion house and 500 acres along with other smaller lots of farms and parcels of land John had bought in the area in the decade before his death, were also sold by public auction. At the same time the Bersham ironworks, by then held leasehold, was sold to Thomas Fitzhugh of Plas Power on whose estate it stood.

Thomas Jones had continued to run the Brymbo estate after his uncle's death as a going concern under the umbrella of the Wilkinson trust – until the peace that followed the Battle of Waterloo produced a dramatic slump in the demand nationwide for cannon and weapons of war. In those difficult times, and perhaps also influenced by the continuing litigation, James Adam had decided to let the Brymbo ironworks in 1818 to John and James Thompson at an annual rental of £1,500, which was increased to almost £10,000 when the other properties and lands, though not Brymbo Hall itself, were included.

That lease was up by the date of the auction, when the three co-heirs, presumably the two girls with their spouses, were all said to be living at the Brymbo Hall.[23] It was at this point that John Wilkinson Jnr and Thomas Jones Wilkinson decided to restart the works, though precisely under what financial arrangement, and whether leasehold or freehold, is not clear. The partnership broke up in 1837 when John Wilkinson Jnr is said to have gone to America. Then in 1840, after a brief stop-gap management by William Rowe, a trusted former employee of John at the Bersham Ironworks, the Brymbo works were sold to a Robert Roy, backed by a Scottish bank and with a brilliant engineer called Robertson involved. It was to usher in a new era of prosperity at Brymbo, but the Wilkinson connection with the place was thus finally ended.

The great ironworks at Bradley, successfully managed for the trust by John's other close relative, William Johnston, throughout the period of devastating litigation, was also sold in the 1830s but continued to produce iron for another hundred years. The Hadley ironworks also produced iron goods and, like Bradley, income for the Wilkinson trust throughout this period until it too was closed down. Hadley was out of production by 1825 and sold out of Wilkinson hands as a small estate of 65 acres in 1831.[24]

In a few brief years, therefore, in the late 1820s and 1830s the Wilkinson empire, until so recently wealthy and powerful and seemingly indestructible, was sold off and dispersed. The ironmaster had gone, his influence broken, his estates wasted. His detractors continued to mock, and perhaps even to relax a little bit. No laudatory biography followed in spite of his importance and nothing much was written about him for the next hundred years. Perhaps he was overtaken by that drive for respectability under the new Victorian morality and became 'that man we don't talk about'.

There is a teasing question, though, which ought to be addressed at the end. If John was such a good judge of character as his management practices suggest, how is it he appointed trustees that could not serve his purpose and prosecute his intentions; and why did he allow them such latitude to buy and sell? Astute business man as he was, he should have known that greed and self-interest would emerge out of these freedoms, even with good men in charge.

Something failed him towards the end. Perhaps it was just a declining body and mind; the onset of Alzheimer's; or maybe his judgement was clouded by that bitterness towards his brother and certain other contemporaries who he believed had betrayed him in his iron-making world. Perhaps it was his divided life between two women, the first a beloved wife but the second a woman he also loved; or maybe his judgement was affected by the satisfaction, even the unexpected joy, of securing an heir, the last of his new children. From the evidence of the will, which is the best evidence we have of his thinking in the last years, he wanted those children to start life well; to have a good, a happy, beginning in the place that had come to mean so much to him. All else was secondary, just so long as it got them into adulthood as educated, thinking, feeling people.

Did it? Perhaps he wanted for them what he knew was important and believed was good for himself, what he described to his loved wife, Mary, and what his lifestyle and his business dealings, certainly to his detractors, seemed repeatedly to contradict:

> a new mixture of fat and lean, sweet and sour or of good and evil … preferable for us upon the whole, to that choice which left entirely to ourselves we might make of having only the former without any seasonings of the latter [25] …

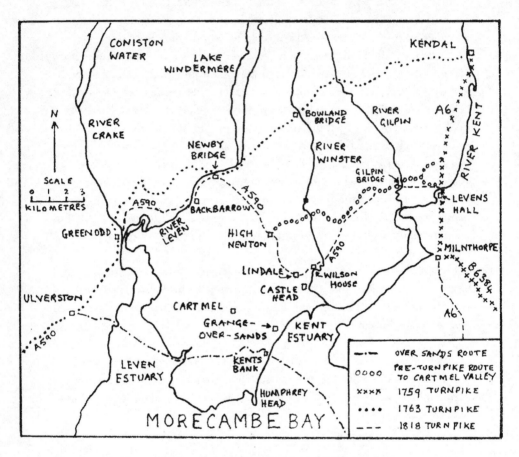

Map showing Morecambe Bay, including the Lindale and Castle Head areas.

REFERENCES

1 Beginnings

1   Janet Butler, archive, Ironbridge Gorge Museum Trust Library, Box 3-1992 10018 33 p. 48.

2   National Library of Wales, NLW Ms. 10823E. Letter (photofacsimile) from Abraham Darby at Madeley Court to Mr William Rolleson, near Kandale, Westmoreland, 12 May 1712.

3   Rawlinson Family Papers, Lancs RO, Preston. DDMc 30/28. Agreement between Backbarrow Company and Isaac Wilkinson, 25 July 1735.

4   Lancaster, J.Y., 'Isaac Wilkinson and the Little Clifton Blast Furnace', *Journal of Cumbria Family History Soc.* (February 2000), p. 14. See also discussion on Isaac's remarriage, beginning of Chapter 9.

5   Davies, Ron, *John Wilkinson* (The Dulston Press, Worcester, 1978).

6   Rawlinson Family Papers, *op. cit.*, as at 3 above.

7   Rawlinson Family Papers, *op. cit.*, DDMc 30/3 and 30/4.

8   *Ibid.*

9   Patent Office, 'Patent No.565, 8 July 1738, Isaac Wilkinson'.

10   *Ibid.*, wording of patent application.

11   Rawlinson Family Papers, *op. cit.*

12   *Ibid.*, Backbarrow Company journals for 1737, p. 462.

13   *Ibid.*, Bar Iron Acc/ts, DDMc 30/12.

14   *Ibid.*, Cast Iron Wares Acc/ts, 27 December 1737, DDMc 30/10.

15   Patent Office, *op. cit.*

16   Patent Office, 'Patent No.713, 12 March 1757, Isaac Wilkinson'.

17   Public Record Office, Ref: C108/135 5402.

18   Nicholson, F. and Axon, E., *The Older Non-Conformity in Kendal* (Titus Wilson, Kendal, 1915).

19   *Ibid.*, letter from Revd Caleb Rotherham to Edward Blackstone of Kendal, September 1735.

20   Dickinson, H.W., *John Wilkinson, Ironmaster* (Hume Kitchen, Ulverston, 1914).

21 Rawlinson Family Papers, *op. cit.*, DDMc 30/36, lease from 1 May 1741 for twenty-one years.

22 *Ibid.*, DDMc 30/6, Backbarrow Company journal, p. 681.

23 *Ibid.*, DDMc 30/11.

24 *Ibid.*, DDMc 30/42.

25 *Ibid.*, DDMc 30/60.

26 Fell, Alfred, *The Early Iron Industry of Furness & District* (first published 1908, reprinted 1968 Frank Cass & Co., Cass Library of Industrial Classics No.21).

27 Rawlinson Family Papers, *op. cit.*, DDMc 30/76, lawyer's opinion in dispute between Lowood Company and wood owners, January 1754.

2 Wilson House

1   Dickinson, H.W., *John Wilkinson, Ironmaster* (Hume Kitchen, Ulverston, 1914).

2   Stockdale, James, *Annals of Cartmel* (first published 1872 as *Annales Caermoelenses*, Michael Moon reprint 1978), p. 210.

3   Dickinson, H.W., *op. cit.*

4   Birmingham Reference Library, Boulton & Watt collection, letters from John Wilkinson to James Watt from Wilson House, 24 August 1778 and 15 November 1778.

5   Stockdale, *op. cit.*, p. 211.

6   *Ibid.*

7   *Ibid.*

8   Lancashire Record Office, Preston, Stockdale Family papers, Letter from John Wilkinson to James Stockdale, 14 July 1787.

9   Stockdale, *op. cit.*, p. 212.

10 Dickinson, *op. cit.*

11 Cumbria Record Office, Kendal, Kirkby Lonsdale Parish Register of Marriages, WPR/19 1754-67.

12 Cumbria Record Office, Kendal, WD/RIG.

13 Ref for Isaac's lease of Bersham furnace from Sir Richard Middleton of Chirk Castle in 1753.

14 Apperley, C.J., *My Life and Times* (E.D. Cuming, 1927), p. 4, p. 219, p. 244. Apperley apparently grew up at Plas Grono after his father, Thomas

Apperley, took over the lease in 1774. See also Challoner, W.H., 'Isaac Wilkinson, Potfounder', *Studies in the Industrial Revolution* (Presnell, L.S. (ed.), 1960).

15 Stockdale, *op. cit.*, pp. 212–13.

3 Bersham – A New Beginning

1   Jacobs, Colin A.J., *The Bersham & Clywedog Industrial Trail* (Bersham Industrial Heritage Centre).
2   Patent Office. Patent No.713, 12 March 1757, Isaac Wilkinson.
3   British Museum BMAM Egerton Mss, 1941 Fol 5–20.
4   Private information from Mrs G.M. Rose, Bristol. See also Palmer, A.N., *John Wilkinson and the Old Bersham Ironworks* (Transactions of the Cymmrodorion Society for 1897–98 (1899)), p. 32.
5   'Yates', *Dictionary of National Biography*.
6   Wedgwood, Hensleigh & Barbara, *The Wedgwood Circle 1730–1897* (UK Publishers Studio Vista, 1980), p. 16. Hensleigh Wedgwood is in direct line of descent from Josiah I, founder of the Wedgwood Dynasty, and had access to unpublished letters and papers for this work.
7   *Ibid.*
8   A detailed and well-documented discussion of this episode can be found in Thomas, Emyr, *Coalbrookdale and the Darbys Published Sessions* (Book Trust (in association with the Ironbridge Gorge Museum Trust), 1999), pp. 50–6.
9   Forester Papers, Shropshire Archives 1224/4/143.
10 *Ibid.*
11 Shropshire Archives – 1396/3 – Lawyers' abstract 22 March 1790.
12 Forester Papers, Shropshire Archives – 1224/4/143.
13 Shropshire Archives – 1396/3 – *op. cit.*
14 Davies, A. Stanley, 'Isaac Wilkinson (*c.*1705–1784) of Bersham, Ironmaster and Inventor' (paper read at the Science Museum, London, 8 February 1950), *Newcomen Society Transactions* 1949–50; quoting Lloyd, John, *Old South Wales Ironworks* (first published London, 1906), and stating the deeds are in a Brecon solicitors office.
15 Original document held at Bersham Industrial Heritage Centre.
16 Taylor, M.S., 'The Plymouth Ironworks', *Glamorgan Historian*, Vol. 5 (1968), p. 186, quoted in Atkinson, M. and Barber, C., 'The Growth &

Decline of the South Wales Iron Industry 1760–1880', *Social Science Monograph*, No.9 (Cardiff, University of Wales Press, 1987).

17   Shropshire Archives – 5001/1/1 21 & 22 and 24/15 a & b.

18   'Lee family of Wroxeter', *Victoria County History of Shropshire*. See also Shropshire Archives – 4044/91 p. 13, 4825, 5001/1/1/4–9, 11–15, 17–22 and 44–5.

19   Forester Papers, *op. cit.* – 1224/7/26.

20   Pee, Ralph, 'The Broseley Home of John Wilkinson', *Wilkinson Society Journal* (1975).

4 Wealth and Acclaim

1   Patent Office. Patent No. 824, 7 February 1765, John Wilkinson.

2   Birmingham Reference Library, Matthew Boulton Collection, Assay Papers 1–66, Letter from John Wilkinson to John Florry, 5 December 1766.

3   *Ibid.*

4   British Library, Ref: EG 1941, Letter Samuel More to Dr Lewis, 12 Mar 1766.

5   See Chapter 6.

6   Challoner, W.H., 'Isaac Wilkinson, Potfounder', in Presnell, L.S. (ed.), *Studies in the Industrial Revolution* (1960) – Quoting from contract with three local master-colliers for obtaining coal from new pit at Cwm Canaid near Merthyr Tydfil.

7   Warrington Reference Library, Priestley Letters, Letter 49, Joseph Priestley to John Wilkinson 19 September 1796.

8   Aikin, J.A., *Description of the Country from Thirty to Forty miles round Manchester* (1795), p. 399, quoted in Challoner, *op. cit.*

9   Dowlais, Plymouth and Cyfarthfa Ironworks. See Lloyd, John, *The Early History of the Old South Wales Ironworks 1760–1840* (1906).

10   Wilkins, Charles, *History of the Iron, Steel, Tinplate & Other Trades of Wales* (1903), p. 42, quoted in Challoner, *op. cit.*

11   For a good synopsis of this litigation see Challoner, *op. cit.*

12   Bristol Reference Library, Sketchley's Directory 1775 – Wilkinson, Isaac, iron founder, 7 Hampton Court.

13   *Gentleman's Magazine*, Vol. 4, Pt. 7, p. 151 (31 January, 1784) – Mr Isaac Wilkinson, formerly an ironmaster at Bersham in Denbighshire.

14   Birmingham Reference Library, Boulton Papers, letter from Matthew Boulton to James Watt, 23 September 1781. Messrs Samuel Walker & Co. Rotherham, Ironmasters, with whom Wilkinson, and Boulton & Watt, did business.

15   See Chapter 2.

16   Documents examined by the author in the winter of 1982–83, courtesy of Mr Norman Kerr, Bookseller, of Cartmel, Cumbria.

17   *Ibid.*

18   Stockdale, James, *Annales Caermoelenses* (1870, Michael Moon reprint 1978 as *Annals of Cartmel*), p. 214.

19   Fell, A., *The Early Iron Industry of Furness and District* (1908, Frank Cass reprint 1968), p. 233.

20   Hoyland, John, 'John Wilkinson – Man of Iron', *The Chartered Mechanical Engineer* (September 1965).

21   Challoner, W.H., 'Translation of Marchant de la Houliere's report to French Government', *Edgar Allen News* (December 1948–January 1949).

22   *Ibid.*

23   *Ibid.*

24   Braid, D., in *Ordnance Society Newsletter*, No. 9 (January) – William Wilkinson's salary from the French government rose quickly to 72,000 livres or £2,500 at the then rate of exchange, approximately twenty times that amount in 1990 values according to Braid.

5 The New Steam Engine

1   Dickinson, H.W. and Jenkins, R., *James Watt and the Steam Engine* (1927, Encore Edition 1981) – for a detailed and well-documented account of the development of the Kinneil experimental engine see Chapter VIII, pp. 93–107.

2   *Ibid.*

3   *Ibid.*, p. 38.

4   Muirhead Papers, Watts Journal September 29th, 1773.

5   *Ibid.*

6   *Ibid.*, undated fragment addressed to Small.

7   Dickinson & Jenkins, *op. cit.*, p. 40 – quoting letter from James Watt to Dr Small, November 1773.

8   *Ibid.*, p. 41.

9 Watt's Public Act Patent No.969. Journal of the House of Commons XXXV, 142, 168, 185, 280, 313, 387.

10 Dickinson & Jenkins, *op. cit.* – for Watt's rehearsal of terms see pp. 43–4.

11 Boulton & Watt collection, Birmingham Reference Library.

12 *Ibid.*, letter from John Wilkinson to James Watt.

13 *Ibid.*, letter from John Wilkinson to James Watt, 5 April 1776.

14 Dickinson & Jenkins, *op. cit.*, p. 40 – quoted from Boulton Papers in Muirhead Collection.

15 Boulton & Watt collection, *op. cit.*, all letters from this source.

16 Boulton Papers, Birmingham Reference Library. Letter from James Watt to Matthew Boulton, 8 July 1778.

17 Samuel More, Journal, held in private hands, *op. cit.*, all details of the itinerary to the Coalbrookdale area are from the 1776 Journal.

18 Boulton & Watt collection, *op. cit.*, letter from John Wilkinson to Mr Watt at Soho, 7 May 1776.

19 Samuel More, Journal, *op. cit.*, – entry for 20 July 1776.

20 Boulton & Watt collection, *op. cit.*, letter from John Wilkinson to Mr James Watt at Soho, 8 September 1776.

6 The Iron Bridge

1 New Willey Company 1759, New Bersham Company 1762 and Bradley Furnace 1772.

2 Shropshire Archives – 6001–3689, the Minute Book of the intended Bridge between Madeley Wood and Benthall 1775–98.

3 Ironbridge Gorge Museum Trust Archive.

4 Minute Book, *op. cit.*

5 *Ibid.*, meeting on 15 May 1776.

6 Ironbridge Gorge Museum Trust Archive, *On Cementitious Architecture as applicable to the Construction of Bridges*, 1832, John White.

7 Minute Book, *op. cit.*, meeting on 17 October 1775.

8 *Ibid.*

9 *Ibid.*

10 *Op. cit.*, entry for 22 January 1776.

11 *Op. cit.*, meeting on 15 May 1776.

12 *Op. cit.*, meeting on 28 July 1776.

13 *Ibid.*

14   *Op. cit.*, meeting on 1 October 1776.

15   *Op. cit.*, meeting on 18 October 1776.

16   Trinder, Barrie, *The Darbys of Coalbrookdale* (Phillimore, 1981 edition), p. 43.

17   Minute Book, *op. cit.*, meeting on 31 March 1777.

18   *Op. cit.*, meeting on 2 May 1777.

19   *Op. cit.*, meeting on 14 July 1777.

20   *Op. cit.*, meeting on 20 October 1777.

21   Shares were assigned as follows: Revd Mr Harries 10, Abraham Darby 15, John Wilkinson 12, Leonard Jennings 10, Samuel Darby 1, Edward Blakeway 2, Mr Farnolds Pritchard 2, John Morris 2, Charles Guest 2, Roger Kynaston 1, John Hartshorne 1, Sergeant Roden 1, John Thursfield 1, John Nicholson 1. N.B. Samuel Darby, Abraham's brother, held 4 shares in the final allocations which is where the discrepancy of 3 shares may be in Thomas Addenbrooke's listing.

22   Minute Book, *op. cit.*, meeting dated 5 February 1779, but likely to be 5 March 1779.

23   Warrington Reference Library, Priestley Letter File, No. 16.

7 The Northern Sanctuary

1   See Chapter 4.

2   Birmingham Reference Library, Boulton & Watt collection, letter from John Wilkinson at Wilson House to James Watt at Redruth, Cornwall, 24 August 1778.

3   Boulton & Watt collection, *op. cit.*, letter from John Wilkinson to James Watt 15 November 1778.

4   Lawyers' abstract among documents examined by the author in winter of 1982–3, courtesy of Mr Norman Kerr, bookseller, of Cartmel, Cumbria.

5   *Ibid.*

6   Stockdale Papers, Lancashire Record Office, Preston, letter from John Wilkinson to Mr Stockdale, Carke, over the sands from Lancaster, 23 December 1778.

7   Boulton & Watt collection, *op. cit.*, letter from John Wilkinson to James Watt, 25 January 1780, copying letter from H. Jones. See also letter from John Wilkinson to Mr Watt, engineer, Birmingham, 24 April 1777.

8 Gilbert White of Selborne, letter to his niece, Miss Molly White 19 October 1778, in Massingham, H.J. (ed.), *The Essential Gilbert White of Selborne* (Breshich & Foss, London, 1983), p. 357.

9 Boulton & Watt collection, *op. cit.*, letter from John Wilkinson to James Watt, 12 September 1779.

10 *Ibid.*, letters from John Wilkinson to James Watt, 18 April 1777 and 14 April 1778.

11 *Ibid.*, letter from John Wilkinson to Boulton & Watt, 6 June 1779.

12 Arret de Conseil of French Government, dated 14 April 1778, quoted in Dickinson, H.W. and Jenkins, R., *James Watt and the Steam Engine* (Encore Editions, London, 1927), p. 48.

13 Stockdale Family Papers, *op. cit.*, letter from John Wilkinson to James Stockdale, 15 June 1779.

14 *Ibid.*, letter from John Wilkinson to James Stockdale, 14 March 1779.

15 Boulton & Watt collection, *op. cit.*, letter from John Wilkinson to James Watt 26 September 1780.

16 Stockdale, James, *Annals of Cartmel* (Michael Moon reprint, 1978), p. 203.

17 Boulton & Watt collection, *op. cit.*, letter from John Wilkinson to James Watt, 26 September 1780.

18 Stockdale, James, *Annals of Cartmel* (Michael Moon reprint, 1978), p. 203.

19 More, Samuel, 1783 Journal, 7 September. Journal verified by Royal Society but held in private hands.

20 *Ibid.*, 1784 Journal, 17 September.

21 *Ibid.*, 1784 Journal, 28 September.

22 *Ibid.*, 1783 Journal, 6 September.

23 Boulton & Watt collection, *op. cit.*, letter from John Wilkinson to James Watt 26 March 1781.

24 *Ibid.*, letter from John Wilkinson to James Watt, 7 May 1781.

25 Samuel More, Journals, *op. cit.*, 30 September 1783.

8 Daughter Mary

1 Cumbria Record Office, Kendal, WPR/19 1751–1812, Kirkby Lonsdale Parish Register of Baptisms and Burial, p. 8, 1756.

2 Unpublished research held in Holy Ghost Fathers archive in Bickley, Kent, involving a co-operation between Father Taylor and Dr W.H. Challoner.

3   Journals of Samuel More, in private hands, entry for 14 July 1776.

4   Birmingham Reference Library, Boulton & Watt collection, Matthew Boulton Papers.

5   Journals of Samuel More, *op. cit.*, 20 September 1784.

6   Lancashire Record Office, Preston, Stockdale Family Papers, letter from Wilkinson to James Stockdale, 1 August 1779.

7   Birmingham Reference Library, *op. cit.*, letter from Wilkinson to James Watt, 21 April 1780.

8   *Ibid.*, letter from Wilkinson to Boulton & Watt, 27 February 1781.

9   *Ibid.*, 4 March 1781.

10   Cumbria Record Office, *op. cit.*, WD/Rig Box 6.

11   Lancashire Record Office, *op. cit.*, letter from Wilkinson to Stockdale, 26 December 1781.

12   Birmingham Reference Library, *op. cit.*, letter from Wilkinson in Brussels to Matthew Boulton, 16 January 1782.

13   *Ibid.*, letter from Wilkinson to Watt, 16 July 1783.

14   *Ibid.*, letter from Wilkinson to Watt, 29 July 1783.

15   Lancashire Record Office, *op. cit.*, letter from Wilkinson to James Stockdale, 29 July 1783.

16   Journals of Samuel More, *op. cit.*, 15 August 1783.

17   Ironbridge Gorge Museum Trust archive, *Memorandum of Richard Reynolds*.

18   Journals of Samuel More, *op. cit.*, 18 August 1783.

19   *Ibid.*, 24 September 1783.

20   Birmingham Reference Library, *op. cit.*, letter from Wilkinson to Watt dated 10, 11 and 23 September 1783.

21   *Ibid.*, letter from Wilkinson to Matthew Boulton, 9 October 1784.

22   *Ibid.*, letter from Wilkinson to Watt, 6 November 1783.

23   Birmingham Reference Library, *op. cit.*, letter from Wilkinson to Boulton, 9 October 1784.

24   Journals of Samuel More, *op. cit.*, 3 October 1784.

25   *Ibid.*, 11 October 1784.

26   *Ibid.*, 12 October 1784.

27   In Rathbone, Hannah Mary (Reynolds' granddaughter), *Memoir of Richard Reynolds* (Chas Gilpin, London, 1852), pp. 145–50, p. 164 (Copy in IGMT Ltd Research Library).

28   Cumbria Record Office, *op. cit.*, WD/Rig Box 6.

29   Shropshire Archives, Market Drayton Church Registers.

30 Cumbria Record Office, *op. cit.*, WD/Rig Box 6.

31 Birmingham Reference Library, op cit., letter from Wilkinson to Mr Watt at Harpurs Hill, Birmingham, from Castle Head, 17 November 1785.

32 *Ibid.*

33 Market Drayton church Registers, *op. cit.*

34 Mary Holbrouke's memorial tablet reads: 'HSE Maria Theophili Houlbrookelle carissma conjux. Mortalia reliquit 16 die Junii AD 1786 susque aetatis 31. I. E. T. S. E Maria infantula E. F.' (I am indebted to my friend and Classics scholar, Roy Jennings of Bristol, for drawing attention to the significance of the meaning of the Latin abbreviations in the inscription 'HSE', or *hic sepulta est* (here is buried). But most interestingly 'I. E. T. S. E.' for *in eodem tumulo sepulta est* (in the same tomb is buried), and 'E. F.', *eodem funere* (at the same burial service).)

35 Birmingham Reference Library, *op. cit.*, letter from Wilkinson to Boulton & Watt, from Castle Head, 14 August 1786.

36 A comprehensive and well-referenced account of Wilkinson's iron boats is available in the *Journal of the Wilkinson Society* (now incorporated in the Broseley Local History Society) No. 15 (1987) the Iron Boat Bicentenary Number, which is entirely given over to Richard Barker's detailed essay 'John Wilkinson and the Early Iron Barges', to whose research I am indebted for what follows.

37 Lancashire Record Office, Preston Stockdale Family Papers, *op. cit.*

38 *Ibid.*

9 Brother William and France

1 Challoner, W.H., 'Isaac Wilkinson, Potfounder', *Studies in the Industrial Revolution*, Pressnell, L.S. (ed.), p. 27.

2 Ironbridge Gorge Museum Trust archive (IGMT) Janet Butler Papers, Box 3, 1992, 10018 33, p. 48, Wilkinson Family Tree.

3 Birmingham Reference Library, Boulton & Watt collection, M-2-12-46.

4 See Chapter 1.

5 IGMT Janet Butler papers, 1992, 10018, Box 1, Folder 1, pp. 1–109 – PRO ref: C 108/135 5402.

6 See Chapter 3.

7 In William Wilkinson's Bills of Complaint to Chancery PRO C 33 490 29 of 17 December 1794 and C 33 490 306 of 16 May 1795.

8   In Isaac Wilkinson's Bill of Complaint to Chançery PRO ref: C 12 856 13 of 26 May 1768.

9   See 2 above.

10   In Isaac Wilkinson's Bill of Complaint, *op. cit.*

11   See Chapter 4.

12   *Ibid.*

13   IGMT Janet Butler papers, Box 1, Folder 4, 1992, 10018 19, PRO ref: AM D3 32 f160.

14   Braid, Douglas, 'The Transfer of John Wilkinson's Technology: William Wilkinson in France', *Ordnance Society Newsletter*, No. 9 (January 1990); see also 'Nantes, 22 September 1788' in Young, Arthur, *Travels in France and Italy during the years 1787, 1788 and 1789* (2nd edition, London, 1794).

15   *Ibid.*

16   Lancashire Record Office, letters to John Wilkinson in Stockdale Family papers, DdHr.

17   Warrington Reference Library, Priestley Letters, Letter 18 10 February 1793, Joseph Priestley at Clapton to John Wilkinson.

18   Birmingham Reference Library, Boulton & Watt collection, *op. cit.*, letters between Boulton, Watt and Wilkinson concerning transport of engines to France for Jary, and Perrier (Paris Waterworks), particularly letters from Wilkinson to Watt dated 31 August and 26 September 1780.

19   Lake, David, *Broseley Local History Society Journal*, No. 23 (2001); Translation of William Wilkinson's survey from original manuscript in French held in archive of L'Academie Francois Bourdon, Le Creusot, France.

20   Braid, Douglas, *op. cit.*

21   Young, Arthur, *op. cit.*

22   *Ibid.*

23   Warrington Reference Library, Priestley letters, *op. cit.*, letter from Dr Priestley to John Wilkinson, 19 September 1796.

24   PRO Chancery records ref: C/33/490/29 of 17 December 1794, transcribed in Janet Butler Papers, IGMT Ltd ref: 1992, 10018 12, pp. 13–42.

25   Birmingham Reference Library, Boulton & Watt collection, letters of Anne Watt, letter to James Watt in London, 28 August 1786.

26   Birmingham Reference Library, Boulton & Watt collection Box 26/17.

27   PRO ref: C 33 490 29.

28   *Ibid.*

10 Disagreement, Dispute and Litigation

1   'Sale of Brymbo', *Aris's Birmingham Gazette*, 20 April 1829.

2   Birmingham Reference Library, Boulton & Watt collection, letter from John Wilkinson to Boulton & Watt, 18 June 1792.

3   Lancashire Record Office, Preston, Stockdale Family Papers, DdHr, letter from Thomas Brockbank in Ulverston to James Stockdale of Cark, 5 May 1794.

4   Shropshire Archives, Shackerley Estate list, letters of Gilbert Gilpin 1781/6/10, GG to John Wilkinson, Bersham, 8 November 1794.

5   Lancashire Record Office, *op. cit.*, Stockdale family papers, letter from John Wilkinson to James Stockdale, 8 April 1795.

6   Birmingham Reference Library, Boulton & Watt collection, *op. cit.*

7   Shropshire Archives, *op. cit.*, Letter GG to William Wilkinson, 9 December 1796.

8   Birmingham Reference Library, Boulton & Watt collection, B&W 12, letter (and enclosures) from William Wilkinson to Boulton & Watt, 28 August 1795.

9   *Ibid.*

10   *Ibid.*

11   IGMT Janet Butler papers, Box 2, 1992, 10018 26, pp. 19–27.

12   Shropshire Archives, *op. cit.*, letter from GG to William Wilkinson at Court near Wrexham, 10 January 1797.

13   *Ibid.*

14   Warrington Reference Library, Priestley letters, *op. cit.*, see letters 43, 44 and 46.

15   *Ibid.*, Letter 50 of 20 October 1796.

16   Birmingham Reference Library, Boulton & Watt collection, letter from William Wilkinson to James Watt Jnr, 16 November 1795.

17   *Ibid.*, letter from James Watt at Brymbo to Matthew Boulton, 8 December 1795.

18   A detailed and closely referenced account of move and counter-move in the industrial espionage surrounding the Pirate Engine supplied to James Stockdale's Cark Cotton Mill can be found in Challoner, W.H., 'The Stockdale Family, The Wilkinson brothers, and the Cotton Mills at Cark-in-Cartmel c. 1782–1800', *Cumberland & Westmoreland Archaeological and Antiquarian Society Papers*, XXIV (1964), p. 356.

19   Lancashire Record Office, Stockdale Family Papers, DDHj, letter from William Wilkinson to James Stockdale at Cark, 5 April 1796.

20 Lancashire Record Office, Stockdale Family Papers, letter from William Wilkinson to James Stockdale, 17 April 1796.

21 Birmingham Reference Library, *op. cit.*, letter from James Watt Snr to William Wilkinson, 23 May 1796.

22 Birmingham Reference Library, *op. cit.*, letter from William Wilkinson to Matthew Boulton, 13 January 1796.

23 Birmingham Reference Library, Boulton & Watt collection, letter from William Wilkinson to Matthew Boulton, 1 July 1796.

24 *Ibid.*

25 *Ibid.*

26 Birmingham Reference Library Boulton & Watt collection, Wilkinson, Box 4, p. 64, Hand-bill printed by Tye, Printer, Wrexham dated 29 October 1796.

27 *Ibid.*

28 *Ibid.*

29 Birmingham Reference Library, Boulton & Watt collection, letter from John Wilkinson to Matthew Boulton, 2 March 1796.

30 Birmingham Reference Library, Matthew Boulton Papers, letters of Anne Watt, Anne Watt to James Watt, 17 September 1785.

31 *Ibid.*, Anne Watt to James Watt, 2 October 1785.

32 *Ibid.*, Anne Watt to James Watt, 11 October 1785.

33 *Ibid.*, Anne Watt to James Watt, 9 November 1785.

34 Birmingham Reference Library, Boulton & Watt collection, 'J. W. Senr notes on the case of B&W v J Wilkinson Novr 1795'.

35 Birmingham Reference Library, *op. cit.*, Matthew Boulton Papers, letter from Boulton to John Wilkinson, 5 [?] November 1796.

36 *Ibid.*, letter from John Wilkinson to Matthew Boulton, 3 December 1796.

37 Birmingham Reference Library, Boulton & Watt collection, B2 Box 20.

38 Birmingham Reference Library, Boulton & Watt collection 'J. W. Senr notes on the case of B & W v J. Wilkinson Novr 1795'.

39 Birmingham Reference Library, Boulton & Watt collection, letter from John Wilkinson to Messrs Boulton & Watt, 16 January 1797.

40 *Ibid.*

41 *Ibid.*

11 Quest for an Heir

1   See Chapter 8.

2   See the epitaph he wrote for himself.

3   Stockdale Family Papers, Lancashire Record Office, Preston, letters from John Wilkinson to James Stockdale, in particular 8 February 1788, 5 December 1788, 25 February 1791, 26 January 1795.

4   Stockdale Family Papers, *op. cit.*, 28 March 1788.

5   Shropshire Archives, Shrewsbury, Littlewood Peace & Lanyon collection, Part 1, Dawley & Stirchley Box 5, 1265/203 & 204.

6   *Ibid.*

7   Shropshire Archives above, Letters of Gilbert Gilpin ref: 1781/6/28.

8   Journals of Samuel More, entry for 11 August 1783.

9   John Randall, *The Wilkinsons* (*Barrow News & Mail*, Reprint 1917), Appendix, pp. 4–5.

10   The Staffordshire historian, Revd Stebbing Shaw, writing in 1800, in Davies, R., 'Thoughts on John Wilkinson and Bradley', *Broseley Local History Society Journal*, No.21 (1999), p. 13.

11   Birmingham Reference Library, Boulton & Watt collection, letter to Matthew Boulton, 26 March 1783.

12   See Hadley, *Victoria History of Shropshire*, Vol. XI (1985).

13   Letter from Thomas Telford to Andrew Little from Shrewbury, 3 November 1793, quoted in Gibb, Sir Alexander, *The Story of Telford: The Rise of Civil Engineering* (London; Alexander Maclehose & Co., 1935), pp. 28–9.

14   Institution of Civil Engineers, Gt. George St, London SW1, Archive ref: T/LO, 24.28.

15   See Chapter 1, reference 12 above – A detailed and well-referenced examination of the project can be found in Gibb, Sir Alexander, *op. cit.*, pp. 45–52.

16   Boulton & Watt collection, *op. cit.*, letter from John Wilkinson to Matthew Boulton, 23 October 1796.

17   Lemuel Francis Abbott (1760–1803), well-regarded portrait painter of late eighteenth-century personalities. His portrait of Nelson is well known. Original portrait of Wilkinson in National Portrait Gallery, cat. 3785, a copy also painted by Abbott held by Ironbridge Gorge Museum Trust.

18   PRO reference for the will of John Wilkinson.

12 Posthumous Rumblings

1 PRO reference for the will of John Wilkinson. N.B. See also Ironbridge Gorge Museum Trust archive, Janet Butler Papers, Box 4, 1992, 10018 87, pp. 136–7.

2 *Ibid.*

3 Journal of Samuel More, entry for 7 September 1783.

4 Castle Head, Westmoreland painted by William Daniell in 1814 and published as an engraving the following year in Ayton, Richard, *A Voyage Round Great Britain*, reproduced in Macleod, Innes, *Sailing on Horseback* (T.C. Farries and Co. Ltd, Dumfries, 1988).

5 Graham Sutton, *North Star* (Collins, 1949), pp. 14–16.

6 Held in Westmoreland County Record Office, Kendal.

7 A well-documented examination of many of these transactions can be found in G.P. Jones, 'The Deeds of Burblethwaite Hall 1561–1828' *Cumberland & Westmoreland Archaeological & Antiquarian Society. Transactions (C&WA&A Soc)*, Vol. 62 (1962), pp. 171–97.

8 See History of Brymbo Works, British Steel Corporation in Ironbridge Gorge Museum Trust archive, Acc. No. A430 1973/230.

9 House of Lords journals, order of 29 July 1819.

10 House of Lords journals, p. 760a, dated 6 June 1823; p. 773b, dated 11 June 1823.

11 On 24 May 1821, Cartmel Register of Marriages, Westmoreland RO, Kendal.

12 Fine levied at Lancaster Assizes 10 March 1824, 5 GO4.

13 Westmoreland Record Office, Kendal, Cartmel Register of Burials, 20 March 1823 and 10 March 1826.

14 Westmoreland Record Office, Kendal, Cartmel Register of Marriages, entry No. 568.

15 Article dated 2 March 1826 annexed to will of John Wilkinson.

16 Private correspondence with Elizabeth Rose (*née* Wilkinson) of Bristol whose father collected a huge amount of information on the Wilkinsons but failed to record his sources.

17 Petition to the Court of Chancery dated 23 July 1825 attached to will of John Wilkinson. (Master's Report dated 23 June 1828, the Order made 5 August 1828.)

18 As well as the Castle Head and Brymbo estates and the extensive ironworks and lands in the Manor and Lordship of Bradley in Staffordshire, it included Wilkinson's Flint estate in the parish of Mold,

his Hadley estate in the parish of Wellington, Shropshire, and his estate in the parish and township of Rotherhithe, Surrey.

19   James Stockdale, *Annals of Cartmel* (originally published 1870).

20   St Paul's church, Lindale, register entry for 16 August 1828, at Westmoreland Record Office, Kendal. Copy with Stockdale Family Papers in Hart Jackson papers, Lancashire Record Office, Preston, Ref DDHj.

21   Hart-Jackson papers, *op. cit.*, ref: DDHj 172, May 7 1870.

22   See reproduction in W.D.A., 'Larger than Life', *Lancashire Life* (August 1975).

23   History of the Brymbo Works, *op. cit.*

24   See Hadley, *Victoria County History of Shropshire*, Vol. XI (1985).

25   Lancashire Record Office, Preston, Hart-Jackson papers, ref DDHj, letter from John Wilkinson to Mary, his wife, 28 March 1788.

INDEX

INDEX